WAR, CLAUSEWITZ AND THE TRINITY

For Mum

War, Clausewitz and the Trinity

THOMAS WALDMAN
University of York, UK

Routledge
Taylor & Francis Group
LONDON AND NEW YORK

First published 2013 by Ashgate Publishing

Published 2016 by Routledge
2 Park Square, Milton Park, Abingdon, Oxon OX14 4RN
605 Third Avenue, New York, NY 10017

Routledge is an imprint of the Taylor & Francis Group, an informa business

British Library Cataloguing in Publication Data
Waldman, Thomas.
War, Clausewitz and the Trinity.
1. Clausewitz, Carl von, 1780–1831 – Philosophy. 2. War (Philosophy) 3. Military art and science.
I. Title
355'.02'092–dc23

Library of Congress Cataloging-in-Publication Data
Waldman, Thomas.
War, Clausewitz and the trinity / by Thomas Waldman.
p. cm.
Includes bibliographical references and index.
ISBN 978-1-4094-5139-6 (hardback)
1. War. 2. Politics and war. 3. Clausewitz, Carl von, 1780–1831. Vom Kriege.
I. Title.
U102.W263 2013
355.02—dc23

2012028078

ISBN 13: 978-1-4094-5139-6 (hbk)

Contents

Foreword

In his essay that accompanies the edition of *On War* edited by Michael Howard and Peter Paret, Bernard Brodie tendered the bold claim on behalf of Carl von Clausewitz that '[h]is is not simply the greatest but the only truly great book on war'. Brodie's judgment has stood the test of some time since it first appeared in 1976. The passage of time and succeeding cohorts of critics have yet to damage the plausibility and authority of *On War*. Indeed, the more carefully the book is examined, the more persuasive is its author – at least, such has been this student's experience. Critics have come and gone, but none have identified serious weakness in *On War*. Given the all too rich strategic history of the past two centuries, and the potency of motivations that have fuelled sceptical or downright hostile assessments of Clausewitz's work, it is obvious that the book has to have extraordinary merit. What is remarkable is that although *On War* plainly has some weaknesses, those flaws are really as nothing compared with the book's strengths. Today, we do not read Clausewitz only as an important 'classic', but also and predominantly because it remains, as Brodie claimed, far and away the most enlightening study of its awful subject that anyone has written. Clausewitz is usefully complemented by Thucydides and Sun Tzu, as well as by Machiavelli and a handful of lesser sages, but his work stands alone in merit, he is not a *primus inter pares*.

Thomas Waldman has written an outstandingly good book about the outstandingly good book that is *On War*. In an obvious sense, Dr. Waldman's mission might be said, rather churlishly, to be an easy one. After all, *On War* is not a hard work to praise. However, this study of Clausewitz reveals, to my admittedly somewhat prejudiced mind, just why *On War* is the greatest book on its subject ever written. Dr. Waldman needs to take no prisoners in his civilised but deadly dismissal of Clausewitz's recent critics. His significant achievement here, though, is not to demonstrate the shallowness of Clausewitz's critics, which is not an over-taxing task, but rather to unpack the argument in *On War* from a perspective that is reasonably novel as well as persuasive (a useful combination).

If the 1990s were dominated by critics of Clausewitz, the 2000s in contrast saw a modest but still substantial canon of good-to-excellent books that revisited Clausewitz's theory and did not find it seriously wanting. Following fairly closely on from useful (and more) works by Antulio Echevarria, Hew Strachan, John Sumida, Hugh Smith, and Beatrice Heuser, to cite the leading authors, Thomas Waldman was presented with more scholarly assistance than an author would like, ideally. It is easier to counterpunch against manifest recent error, than it is to risk appearing to say 'me too'. This book does not do that. While Dr. Waldman does say, in effect, 'me too' with the legion of Clausewitz's admirers, he set himself the daunting task of penetrating the unifying core of the argument in *On War*, Clausewitz's 'wondrous trinity'.

This study is the most successful effort I have seen in explaining the nature of war as a true *gestalt*. Dr. Waldman reveals convincingly why and how the superficially reductionist primary and secondary trinities (passion, chance, and reason; people, army and its commander, and policy) are nothing of the kind, when they are understood properly. To the best of my knowledge, no-one has unpacked Clausewitz's theory of war more convincingly than does Dr. Waldman here. With great skill and amazingly in accessible prose, Dr. Waldman presents the subject of war in Clausewitz's trinitarian terms with all the respect for complexity, nuance, ambiguity, and uncertainty that the master's treatment implies and sometimes recognises. Clausewitz intended to present a holistic understanding of war, with distinctly porous categories in its working parts. His primary trinity is better explained with a Venn diagram than it is with the triangular geometry that is so expedient for PowerPoint presentation.

Because *On War* is an educational work, often it seems irrelevant to the problem-solvers in the military profession. As Beatrice Heuser has explained persuasively, it is a rarity for soldiers to be very interested in the nature of war, and therefore by logical extension in theory to reveal that nature. Clausewitz provides the most convincing nuanced answer by far to questions that appear a luxury to practical people. Such people typically cannot study *On War*, even if they had the time, and derive from it the kind of assistance that they believe they need in order to answer the live problems of the here and now. But, in order to function effectively as soldiers or civilian officials there is acute need for sound education that explains the nature and working of statecraft, war, and strategy. Clausewitz was admirably explicit in denying an intention to provide a 'sort of *manual* for action'. *On War* can provide the understanding of its subject that allows those who are educable to be able to design and execute answers that might be good enough to meet the strategic problems that they face. Of course, theory can only educate, but education can prove a great enabler of practical effectiveness.

Dr. Waldman's explanation of Clausewitz's theory of war is a major contribution to the provision of better theory for better practice. It is true that most of that better theory was always lurking in the pages of *On War*, but the Prussian was often subtle and nuanced to a fault. *War, Clausewitz and the Trinity* joins the small honour-roll of first rate explanations of Clausewitzian theory.

Colin S. Gray
Professor of International Politics and Strategic Studies,
University of Reading, England

Preface

In November 2007, I travelled to a remote Afghan town, close to the Tajik border in Badakhshan Province, called Shahri Buzurg (meaning 'Great City' – it was anything but that, however the people were incredibly warm and welcoming). The landscape around the town was dotted with the rusting remnants of the war against the Soviets in the 1980s. A meeting was arranged for me with a veteran of that struggle – a proud, handsome and rosy-cheeked former mujahedeen commander. At the time, I was about half-way through writing the present work and was therefore keen to learn whether the commander had ever heard of Clausewitz. It turned out that he had not only heard of Clausewitz, but had read *On War* in some depth and had consulted it often during the fighting against Soviet forces.

A number of commentators have, in recent years, proclaimed that the rise of insurgency and other forms of irregular, asymmetric conflict have rendered Clausewitz's insights – forged in an age dominated by inter-state warfare – no longer useful or relevant. That an insurgent commander placed such faith in a purportedly obsolete work therefore represents something of a paradox. But, moreover, this anecdote is perhaps the first clue as to why those denying Clausewitz's contemporary relevance are mistaken. The rest of this book is essentially devoted to explaining why the apparent paradox is in fact false. It seeks to put analytical flesh on the central contention that many of Clausewitz's most important ideas do still apply.

*

I have accumulated many debts in writing this book. Aside from those owed financially to NatWest Bank and my brother, most are debts of gratitude. This book represents over five years of study, but its genesis goes back further: my interest in history and war was first sparked thanks to the inspirational and engaging teaching of Mr David Jackson at Lady Manners School in Derbyshire, who set me on the trajectory that ultimately led to me writing this book.

I would like to thank Christopher Bassford and Antulio Echevarria II – two leading interpreters of Clausewitz – for the time they put aside to talk to me: I learnt a great deal in those discussions. Two extremely able friends – Ben Richardson and Kathryn Rzeszut – were generous enough to read the manuscript and provide excellent comments. Also, sincere thanks go to Kirstin Howgate, Margaret Younger and Sadie Copley-May at Ashgate, who have been so patient, kind and helpful throughout the publishing process. Many thanks to Warren Chin, Editor of *Defence Studies*, and Robert H. Taylor, Editor of *Parameters*, for publishing earlier articles associated with this book. I am especially grateful to the modern master of strategic thought, Colin S. Gray, who kindly agreed to add his imprimatur to the book and write such an excellent, apposite Foreword.

Many thanks to Duncan Kelly, Rhiannon Vickers and Nick Rennger who provided valuable guidance at various stages of this project. Crucially, I am incredibly lucky to have such a wonderful friend and mentor in Caroline Kennedy – one of the world's most perspicacious thinkers on war and security – and regarding whom I cannot help but be reminded of Clausewitz's observation that 'a *wise teacher* guides and stimulates a young man's intellectual development, but is careful not to lead him by the hand for the rest of his life'.

Closer to home, many fine friends have contributed inestimably to this book in ways they might not realise, in particular: James Buffoni, Cameron and Jessie Christie, Ben Collard, Nick Corney, Sam Fletcher, Ben McGregor, Andy Mumford, Dom Reeve, John and Rachel Roberts, Euan Robinson, Jess and Ruth Townend, Nick Turner and Adam White. I am particularly indebted to Olga Tregubova for all her kindness and support, not to mention grammatical guidance! The wider Snell tribe – Andy and Josie (and Mugs!); Matt, Suzanne, Amy, Joe and Jacob; Dave, Kal, Isaac and Sam; Jo and Paul – are a constant and invaluable source of laughs, wisdom, encouragement and precisely apportioned deprecation! Andy deserves special mention for keeping me sane during the most difficult stages of the research and always providing such sage advice.

I am so lucky to have such a dependable, fun and wonderfully mad family: Anne, Phil and Will (and Bertie!); Janie, Jack and Dan the Man; Peter, Jane, Andrew and Annabel; Tom and Jo; Anna and Rob; Angie and Brett; and Pep – each a true legend and inspiration in their own way. My brother, Matt Waldman, who has spent years dealing with the very real intersection of war and politics in Afghanistan: I cannot thank you enough for all your encouragement and wise direction. There is so much of our amazing father, Fred, in you. He would be so proud of your incredible accomplishments, and I hope he would approve of this book!

Mum and Steve – what more can I say, this book simply wouldn't have been possible without your love, patience and incredibly selfless support. Mum, I've dedicated the book to you – I don't know of a more generous and caring person; thank you for *everything*.

Thomas Waldman
York, 2012

War is more than a true chameleon that slightly adapts its characteristics to the given case. As a total phenomenon its dominant tendencies always make war a paradoxical trinity – composed of primordial violence, hatred, and enmity, which are to be regarded as a blind natural force; of the play of chance and probability within which the creative spirit is free to roam; and of its element of subordination, as an instrument of policy, which makes it subject to reason alone.

The first of these three aspects mainly concerns the people; the second the commander and his army; the third the government. The passions that are to be kindled in war must already be inherent in the people; the scope which the play of courage and talent will enjoy in the realm of probability and chance depends on the particular character of the commander and the army; but the political aims are the business of government alone.

These three tendencies are like three different codes of law, deep-rooted in their subject and yet variable in their relationship to one another. A theory that ignores any one of them or seeks to fix an arbitrary relationship between them would conflict with reality to such an extent that for this reason alone it would be totally useless.

Our task therefore is to develop a theory that maintains a balance between these three tendencies, like an object suspended between three magnets.[1]

Carl von Clausewitz, *On War*, trans. and ed. Michael Howard and Peter Paret (New York: Everyman's Library, 1993), p. 101.

Chapter 1
Introduction

Writing in the early nineteenth century, Carl von Clausewitz (1780–1831), a Prussian general, produced what is perhaps the most famous book on the nature of war. Indeed, Bernard Brodie wrote that his masterpiece, *On War*, 'is not simply the greatest but the only true great book on war'.[1] Harry Summers believed that in military science it 'is still the seminal work'.[2] Christopher Coker describes Clausewitz as 'the greatest phenomenologist of war'.[3] Despite such acclaim, it has been observed that his book is often mentioned, occasionally quoted, but little read.[4] As a result, it has also been frequently misunderstood. Studies employing purportedly Clausewitzian insights that demonstrate little evidence of deep familiarity with the text are commonplace.

Clausewitz is mentioned ubiquitously in strategic studies and military historical literature and often in an extremely instrumental fashion. His ideas are used in a selective and cursory manner, so that he appears everywhere but often only in passing. The real Clausewitz is lost in a mass of fleeting references, most of which are misleading or do not fully convey the complexity of his arguments. Many modern popular works deploy supposedly Clausewitzian ideas that are almost unrecognisable to anyone closely acquainted with his work or that present such crude, often derivative, representations as to appear almost satirical. Most striking is the disproportion between the towering status of his name and how little he appears to have been studied in any real depth. As is often the fate of great works, they are scavenged for ideas to lend gravitas to spuriously related arguments.

It has been written that, in order to avoid such endless misunderstanding, *On War* 'has to be studied repeatedly, seriously, and in depth'.[5] The irony that this was written by one of Clausewitz's most mistaken interpreters should not

---

1 Bernard Brodie, 'The Continuing Relevance of *On War*', in Carl von Clausewitz, *On War*, trans. and ed. Michael Howard and Peter Paret (New York: Everyman's Library, 1993), p. 58. Hereafter, all references to *On War* will refer to this edition, unless otherwise stated.

2 Harry Summers, *On Strategy: The Vietnam War in Context* (Honolulu: University Press of the Pacific, 2003), p. 4.

3 Christopher Coker, *Waging War Without Warriors? The Changing Culture of Military Conflict* (London: Lynne Rienner, 2002), p. 54.

4 Christopher Bellamy, *Knights in White Armour: The New Art of War and Peace* (London: Pimlico, 1998), p. 25.

5 Martin van Creveld, 'The Eternal Clausewitz', in Michael I. Handel (ed.), *Clausewitz and Modern Strategy* (Abingdon: Routledge, 2005), p. 36. Creveld's analysis and interpretation in this article is clear and accurate – it is his later work that misrepresents Clausewitz's ideas.

detract from the wisdom of the injunction. More importantly, Clausewitz must be studied in context because, 'Every theory that outlasts its creator tends to be reinterpreted unhistorically ... Clausewitz's writings have suffered the attendant distortions more than most'.[6] Of course, interpretations are inevitably subjective, yet this does not prevent us from recognising where some are superior to others. The prominent sources of forgivable misinterpretations are well-known: the unfinished text, the complexity of the ideas, the use of outdated concepts and problems associated with translation. The obscurity of his work has thus allowed for radically divergent readings.[7] The only means of overcoming these pitfalls is through a detailed analysis of *On War* and Clausewitz's wider work, attempting to understand the man himself and situating his ideas in the time they were written, as well as building on reliable secondary studies.

To guard against mistaken readings of *On War*, a key aim of the book is to provide a thorough interpretation of Clausewitz's central theoretical device: his 'remarkable trinity' of passion, chance and policy. The trinity can be considered the apotheosis of his theorising. His brief description of the trinity comes at the end of Book 1, Chapter 1 – the only chapter we know he considered complete[8] – and is presented under the subheading, 'The Consequences for Theory'. In this respect, the trinity represents his central analytical framework for comprehending the nature of war. It can be viewed as *On War* condensed into five paragraphs: it 'ties all of Clausewitz's many ideas together and binds them into a meaningful whole'.[9] Read in isolation, the trinity can and has led to as much confusion as clarity. On first reading it can appear decidedly simplistic, imprecise and perhaps even odd. As Hugh Smith observes, though Clausewitz 'nowhere discusses passion, reason, and chance at length, these elements permeate his entire work'.[10] It might be more accurate to state that he does discuss those elements at considerable length, but rarely explicitly in direct reference to the trinity itself.

It should also be noted that the trinity might appear to some as distinctly cold and analytical – a description of the nature of war which fails to reflect the awfulness of the subject. Yet, Clausewitz never lost sight of the 'horrible reality of war'.[11] He certainly was no militarist, even if he saw war as an occasional necessity and, moreover, one which may entail great bloodshed. War, he believed,

---

6 Peter Paret, *Clausewitz and the State: The Man, His Theories, and His Times* (Princeton: Princeton University Press, 1985), p. 8.

7 Azar Gat, *A History of Military Thought: From the Enlightenment to the Cold War* (Oxford: Oxford University Press, 2001), p. 202.

8 He states that, 'The first chapter of Book One alone I regard as finished'. Clausewitz, 'Unfinished note, presumably written in 1830', in *On War*, p. 79.

9 Christopher Bassford, 'The Primacy of Policy and the Trinity in Clausewitz's Mature Thought', in Hew Strachan and Andreas Herberg-Rothe, *Clausewitz in the Twenty-First Century* (Oxford: Oxford University Press, 2007), p. 75.

10 Hugh Smith, *On Clausewitz* (New York: Palgrave, 2005), p. 116.

11 Brian Bond, *The Pursuit of Victory: From Napoleon to Saddam Hussein* (Oxford: Oxford University Press, 2006), p. 53.

'is no pastime; it is no mere joy in daring and winning, no place for irresponsible enthusiasts. It is a serious means to a serious end'.[12] Nevertheless, he stressed that, 'It would be futile – even wrong – to try and shut one's eyes to what war really is from sheer distress at its brutality'.[13] Clausewitz was not deluded about the reality of his subject, but war, as only a soldier could understand, is more ambiguous in its meaning and impressions than those who have not experienced it may believe. He had witnessed the heart-rending crossing of the Berezina by Napoleon's retreating armies and was moved to write that, 'If my feelings had not been hardened it would have sent me mad'.[14] War could also be intensely enjoyable. Clausewitz often wrote to his wife describing his great excitement regarding impending battles.[15] In sum, war for Clausewitz was an emotionally and morally complex phenomenon, at once dreadful, necessary and exhilarating.

Who was Carl von Clausewitz?

Adventure, learning and love: 1780–1806

Carl Phillip Gottlieb von Clausewitz was born in June 1780 in Burg, seventy miles southwest of Berlin. His father was a retired lieutenant from the Seven Years War who had barely retained his claim to noble status. Clausewitz grew up in a house frequented by officers and received a fairly mediocre provincial education. Just prior to his twelfth birthday he was accepted into the 34th Infantry Regiment commanded by the brother of the late Frederick the Great, Prince Ferdinand. After missing the momentous cannonade at Valmy in 1792, within months he enthusiastically received his first taste of war when his regiment was deployed to take control of the Rhine from French revolutionary forces in early 1793. Clausewitz took part in the series of operations that culminated in the liberation of the strategically important garrison town of Mainz. The army then marched south to the Vosges Mountains where, due to being 'fragmented into tendrils of outposts and detachments', it conducted a series of limited raids and ambushes until the Peace of Basel ended hostilities between France and Prussia in 1795.[16]

These experiences undoubtedly 'provided the impulse … to his early studies and his first theoretical speculations' which were given time to germinate during the time Clausewitz spent garrisoned in the mundane town of Neuruppin between 1795 and 1801.[17] Despite the somewhat prosaic nature of this tour of duty,

12 *On War*, I, 1, § 23, p. 98.

13 *On War*, I, 1, § 3, p. 84.

14 Quoted in Andreas Herberg-Rothe, *Clausewitz's Puzzle: The Political Theory of War* (Oxford: Oxford University Press, 2007), p. 161.

15 Roger Parkinson, *Clausewitz: A Biography* (New York: Stein and Day, 1979), p. 58.

16 Paret, *Clausewitz and the State*, p. 30.

17 Paret, *Clausewitz and the State*, p. 35.

Clausewitz nevertheless benefited from his intimate experience of Prussian military institutions and the pedagogical reforms and enlarged curriculum introduced by the regimental commander, Freidrich von Tschammer.

It was not until Clausewitz's admission to the Berlin Institute for Young Officers in the summer of 1801 that he could effectively begin to expand his education. The institution was under the directorship of the Hanoverian, reformist minded, Gerhard von Scharnhorst who soon became one of Clausewitz's closest friends. Scharnhorst emphasised the development of the students' intellect, powers of judgement and independence of thought, while not forcing on them any particular theory of war. Here, Clausewitz became acquainted with some of the fundamental issues he would later expand on, such as the relationship between theory and practice and the importance of history. (Also, in December 1803, Clausewitz met his future wife, Marie von Brühl, lady-in-waiting to the Queen Mother). Initially struggling to keep up in lectures, he graduated at the head of his class in the spring of 1804 and secured a regular appointment as adjutant to Prince August, bringing him into increasing contact with the imperial court, yet usually as something of an outsider. After much fumbled diplomacy and strategic manoeuvring, Frederick William III's Prussia again found itself on a collision course with France under Emperor Bonaparte, and as Clausewitz prepared to go to war he wrote prophetically that Prussia's situation was 'infinitely wretched'.[18] In September 1806 his regiment began its journey to the battlefield of Auerstedt.

Soldier, teacher and reformer: 1806–1815

Clausewitz's introduction to the Napoleonic form of warfare was a dramatic one. Positioned on the right flank of the Prussian army at the Battle of Auerstedt, in the midst of the fighting he had to take charge of an entire battalion, managing to command a reasonably orderly fighting retreat.[19] Prince August and Clausewitz were cut off from the main force, captured by French troops and subsequently transferred to France where they were interned under 'easy conditions' until the treaty of Tilsit, in July 1807, permitted their repatriation. Before returning to Germany they spent a number of intriguing weeks at the Château de Coppet by Lake Geneva, exile home of Napoleon's great literary foe, Madame de Stäel, whom Clausewitz befriended along with other members of her 'camp' such as August Wilhelm Schlegel and Benjamin Constant. Between Clausewitz's return to Prussia and the war of 1812, Clausewitz became an integral member of the Military Reorganisation Commission headed by Scharnhorst at the new War Department, charged with modernising the army. The reforms entailed not only measures internal to military institutions, but changes which would have far-reaching implications for Prussian society; most notably plans for ending the nobility's monopoly on officer corps commissions and the introduction of universal military service. By

18 Paret, *Clausewitz and the State*, p. 118.

19 This was the only occasion of Clausewitz's direct command of troops in battle.

the end of this period, Clausewitz had risen to become head of Scharnhorst's office and was appointed lecturer at the new *Allgemeine Kreigsschule* where he taught a class on 'Small Wars', became military tutor to the Crown Prince and was placed on the board of the commission responsible for drafting tactical and operational guidelines for the infantry and cavalry.

Clausewitz's intellectual endeavours were again interrupted when the battlefield beckoned for the third time in 1812. Outraged by Prussia's submission to Napoleon's designs, Clausewitz resigned his commission so he could serve with the Russian army.[20] In this capacity, he took part in the dramatic Russian retreat including the battles of Smolensk, Borodino and the fall of Moscow in 1812 – all movingly depicted in Tolstoy's *War and Peace*, in which Clausewitz briefly appears.[21] During the subsequent reversal of fortunes and Napoleon's retreat over the winter, he played an important diplomatic role in the Convention of Tauroggen, which ensured the defection of General von Yorck's Prussian forces that were fighting with the *Grand Armée*. Subsequently, Clausewitz and Yorck returned to East Prussia and coordinated the raising of a militia of 20,000 men, the *Landwehr*: an act that helped persuade Frederick William III to finally declare war on France in March 1813. Still in Russian uniform, due to the King's refusal to readmit him to the Prussian army, Clausewitz served as Scharnhorst's assistant at the Battle of Grossgörschen in May where the latter was fatally wounded. During the Leipzig campaign, which drove Napoleon out of Germany, Clausewitz was far from the main theatre of operations near the Baltic coast serving as chief of staff to the Russo-German Legion. Only in April 1814 did he finally receive a commission as colonel in the Prussian infantry. During the 'Hundred Days' of Napoleon's return in 1815, Clausewitz served as chief of staff to the 3rd Corps, which fought a difficult holding battle at Wavre and kept Marshal Grouchy from reinforcing Napoleon at Waterloo.

Reflection, study and writing: 1815–1831

After the war, Clausewitz was assigned to General August von Gneisenau's headquarters at Coblenz on the Rhine as chief of staff. From then on, he devoted much of his spare time to intensive research and writing, including early drafts of *On War*. In 1818 he somewhat reluctantly accepted appointment as director of the Berlin War Academy, which came with his promotion to major-general. Clausewitz remained in Berlin until a new appointment as commanding officer of the 3rd Inspection of the Artillery entailed his move to Silesia in 1830. However, before having time to settle there, upheavals spread throughout Europe and prompted the

20 It is interesting to note that this meant Clausewitz would not only be fighting against Prussian soldiers, but against his two brothers who remained in that army. Parkinson, *Clausewitz*, p. 136.

21 Leo Tolstoy, *War and Peace*, trans. Rosemary Edmonds (London: Penguin, 1982), p. 920.

mobilisation of part of the Prussian army, primarily in response to the Warsaw insurrection in November. After initially being stationed in Berlin, in March 1831 Clausewitz left for Posen with his friend Gneisenau, then commander-in-chief of the army. An encroaching cholera epidemic from the east meant Clausewitz was entrusted with organising a *cordon sanitaire* on the Prussian frontier and, after the headquarters at Posen was disbanded, he returned to Breslau on 7 November and was joined by Marie two days later. They were able to spend eight days together before Clausewitz contracted cholera and died on the evening of the 16th at the age of fifty-one. The book he had been working on prior to his posting remained unfinished.

Introducing the trinity

So, what exactly is the trinity? It is essentially a succinct expression of ideas composed of a number of central concepts held together by a deeper unifying and integrative logic. Christopher Bassford describes it as an 'amazingly compressed summation of reality'.[22] In formulating the trinity, Clausewitz simply wanted to argue that war is made up of three central elements or 'dominant tendencies'. So, war is,

> composed of primordial violence, hatred, and enmity, which are to be regarded as a blind natural force; of the play of chance and probability within which the creative spirit is free to roam; and of its element of subordination, as an instrument of policy, which makes it subject to reason alone.[23]

These tendencies have been variously condensed into short-hand versions such as 'violence, chance and politics', 'hostility, chance and purpose'[24] or even 'irrational, non-rational and rational factors'.[25] In this book they will generally although not exclusively be referred to as passion, chance and policy (or politics). This set of three elements constitutes the 'primary' or 'objective' trinity. Each of the three tendencies of the primary trinity, Clausewitz explains, is *mainly* manifested in a corresponding subject within society: respectively, the people, the commander and his army, and the government. This set of three elements constitutes the 'secondary' or 'subjective' trinity. In addition to these two central conceptual levels is the often

---

22 Christopher Bassford, 'Tip-Toe Through the Trinity or The Strange Persistence of Trinitarian Warfare'. Available online at: <http://www.clausewitz.com/CWZHOME/Trinity/Trinity8.htm>, retrieved 14 March 2008.

23 *On War*, I, 1, § 28, p. 101.

24 Antulio Echevarria, *Clausewitz and Contemporary War* (Oxford: Oxford University Press, 2007), p. 55.

25 Christopher Bassford and Edward J. Villacres, 'Reclaiming the Clausewitzian Trinity', *Parameters*, Autumn, 1995.

overlooked level of context (not itself comprising a trinity).[26] As we will show, this partially hidden level of the holistic framework serves a vital theoretical role with regard to the way it explains change at the secondary level. Failure to recognise the importance of contextual change can seriously detract from the intended flexibility of the trinity. Context might usefully be considered a 'tertiary' level. So, pulling these strands together, the entire trinitarian framework is composed of three levels:

- *Primary:* passion, chance and policy.
- *Secondary:* people, commander and army, and government.[27]
- *Tertiary:* context.

To avoid confusion, it is important to note how these levels relate to the structure of the book. The foundations of the trinity, discussed below, and Chapter 2, which examines the theoretical foundations of Clausewitz's approach, do not themselves constitute levels in the framework nor relate specifically to any particular level: they underlie the entire framework. Chapter 3 explores the tertiary level of context. The primary tendencies are the principal subject of Chapters 4, 5 and 6. Chapter 7 draws the analysis together, seeking to reveal how all the levels relate to one another.

The foundations of the trinity

The trinity, although complete in itself as a description of the nature of war, is built on a number of central suppositions. These are not explicitly stated in the paragraphs in which the trinity is presented, but are deemed by Clausewitz as fundamental to an understanding of war and are examined in other parts of *On War*. Three stand out: the human element, the interactive dynamic of war and the centrality of fighting. These three areas emphasise, respectively, the importance of maintaining attention on war's ever-present psychological, sociological and material dimensions.

A human social activity

Underlying all of Clausewitz's thoughts on war is the human element – the behaviour and interaction of humans in their social environment. Of all of war's dimensions, the human element is the most enduring. Clausewitz is ever at pains to emphasise this essential, yet easily overlooked reality. Of course, he was well aware that humans throughout history have possessed radically different preconceptions,

26 Context is not usefully delineated into three areas corresponding to the trinity. It is, as explained in Chapter 3, a much more diffuse and complex idea.

27 Chapter 7 will demonstrate why analogues (such as popular base, fighting forces and political leadership) of these groups may be preferred.

attitudes and views of the world shaped by such things as culture, religion and social norms. Nevertheless, he believed that underlying these forms there has always been, and always will be, a common nature to man from which derives the 'human condition', which can help account for noticeable continuities in the behaviour of societies over time and across geographical and cultural divides.

This perspective is easily overlooked because it is war's material character that is perhaps most striking to the observer. Theorists prior to and since Clausewitz have failed to adequately ground their works in the reality that it is a human-centric activity. In modern times, this has been exacerbated by the extent to which technology – with its apotheosis in nuclear weapons, information technology and robotics – appears to dominate the conduct of war. The effect can be blinding in its capacity to divert attention from the basic fact that it is still humans that ultimately wield and employ such technology. Following Clausewitz, Colin Gray reminds us that 'the human element is the greatest of continuities in the long bloody history of warfare'[28] and as MacGregor Knox notes, it is ultimately 'individuals – with their ambitions, vanities and quirks – [that] make strategy'.[29] More insistent still is the remark by North Vietnamese General Vo Nguyen Giap: 'In war there are two factors – human beings and weapons. Ultimately, though, human beings are the decisive factor. Human beings! Human beings!'[30]

Emphasising 'the human' in war and politics places Clausewitz in a tradition of thought most notably represented by Thucydides and Machiavelli. Thucydides hoped that his work would be 'judged useful by those who want to understand clearly the events which happened in the past and which (human nature being what it is) will, at some time or other and in much the same ways, be repeated in the future'.[31] Machiavelli, who rediscovered the human element in war following centuries of medieval providentialism,[32] stated that,

> Everything that happens in the world at any time has genuine resemblance to what happened in ancient times. This is due to the fact that the agents who bring such things about are men, and that men have, and always have had, the same passions, whence it necessarily comes about that the same effects are produced.[33]

---

28 Colin Gray, *Another Bloody Century: Future Warfare* (London: Weidenfeld & Nicolson, 2005), p. 376.

29 MacGregor Knox, 'Conclusion: Continuity and Revolution in the Making of Strategy', in Williamson Murray, MacGregor Knox, and Alvin Bernstein, (eds), *The Making of Strategy: Rulers, States and War* (Cambridge: Cambridge University Press, 2007), p. 645.

30 Quoted in Stanley Karnow, *Vietnam: A History* (London: Pimlico, 1994), p. 21.

31 Thucydides, *History of the Peloponnesian War*, trans. Rex Warner (London: Penguin, 1972), p. 48.

32 Coker, *Waging War*, p. 48.

33 Niccolò Machiavelli, *The Discourses*, trans. Leslie J. Walker (London: Penguin, 2003), p. 517.

As with Thucydides and Machiavelli, the implications derived from the human element in war are rarely spelt out directly by Clausewitz, but allowed to reveal themselves throughout the text. They pervade all aspects of the trinity in differing ways.

The focus on the human factor also explains Clausewitz's persistent concern with psychological and moral aspects of war, not to mention the influence of personalities.[34] In his historical works, Clausewitz was often at pains to delve into the unique psychology of individual commanders and, as Thucydides had done, of ordinary soldiers, as this could explain much about the conduct and course of war.[35] It is this fact which helps explain the trinity's profound realism, anchored as it is in the underlying nature of those who conduct it. Furthermore, Clausewitz recognised that not only is war a human activity, it is also always a human *social* activity; it is waged collectively – as Gray notes, 'individuals, as individuals, do not wage war'.[36]

Interaction – the duel

A saying has it that, 'It takes two sides to make war. It only takes one side to make a massacre'. Likewise, Clausewitz was adamant that war is never simply the unilateral use of violence. Clausewitz perceptively, and somewhat counterintuitively, held that defence is the first true act of war, because without it the use of force is conquest without resistance:

> War serves the purpose of the defense more than that of the aggressor. It is only aggression that calls forth defense, and war along with it. The aggressor is always peace-loving (as Bonaparte always claimed to be); he would prefer to take over our country unopposed. To prevent his doing so one must be willing to make war and be prepared for it.[37]

Resistance thus lies at the heart of the idea of war and 'total nonresistance would be no war at all'.[38] War, he describes, is comparable to a duel; it is violence met by violence. Yet, as Alan Beyerchen has noted, the analogy is more complex than the image of an aristocratic duel. The term used by Clausewitz is *zweikampf*, that literally means 'two-struggle' and which he illustrates through the example of two wrestlers: this form of struggle is altogether unpredictable, 'generating positions

---

34 *On War*, I, 2, p. 107.

35 In his 'Observations on Prussia in Her Great Catastrophe', he devotes an entire section to profiles of the protagonists. Carl von Clausewitz, *Historical and Political Writings*, ed. and trans. Peter Paret and Daniel Moran (New Jersey: Princeton University Press, 1992), pp. 42–63.

36 Gray, *Bloody Century*, p. 378.

37 *On War*, VI, 5, p. 444.

38 *On War*, I, 1, § 4, p. 86.

and shapes that neither could possibly create alone'.[39] It becomes clear when reading *On War* that it is this complex interaction that accounts for so much that is unique about the phenomenon. War is a violent conflict of wills, a 'collision'[40] or 'conflict of living forces'[41] which, on interaction, creates new political and military realties that neither side fully controls. As we will see, this has important implications for all tendencies of the trinity.

Clausewitz criticised earlier thinkers for their underestimation of this point. They developed dogmatic, prescriptive theories which purportedly offered near certain success if followed, forgetting that any prescriptive theory inevitably runs up against an unavoidable problem: what happens if the enemy adheres to the same principles? In such cases, surely the supposedly war-winning principles lose all meaning? Clausewitz wanted to emphasise that, 'In war, the will is directed at an animate object that *reacts*'.[42] For the practitioner this is essential because, 'war is not waged against an abstract enemy, but against a real one who must always be kept in mind'.[43]

The centrality of fighting

Contrary to what his famous dictum – 'war is merely the continuation of policy by other means'[44] – might suggest, Clausewitz held that war must ultimately be defined by its basic means, not by its purpose. For him, the means of war is fighting or combat: essentially, organised violence. War is an act of force. The possible political contexts and purposes of war can be endlessly varied, but at its heart is the act of physical violence. This should not be seen as detracting from the importance of the political perspective, but Clausewitz was at pains to establish what distinguished his subject from any other.

Echevarria describes fighting as the central feature of the Clausewitzian universe of war, just as Copernicus described the actual universe in terms of his heliocentric system.[45] Clausewitz is at times brutally direct in his articulation of this point to ensure there is no confusion: 'War is a clash between major interests, which is resolved by bloodshed – that is the only way it differs from other conflicts'.[46] Even where political factors appear to be prominent during war, often

39 Alan Beyerchen, 'Clausewitz, Nonlinearity and the Importance of Imagery'. Available online at: <http://www.clausewitz.com/readings/Beyerchen/BeyerschenNonlinearity2.pdf#zoom=100>, retrieved 20 January 2012.

40 *On War*, I, 1, § 4, p. 86.

41 *On War*, II, 3, p. 174.

42 *On War*, II, 3, p. 174.

43 *On War*, II, 5, p. 187.

44 *On War*, I, 1, § 24, p. 99.

45 Echevarria, *Clausewitz*, p. 3.

46 *On War*, II, 3, p. 173.

their importance rests on expected military implications.[47] This might apply due to a changed calculus of relative strength, the resultant military superiority gained through new allies and so forth.

Clausewitz's emphasis on fighting is not to say that he believed decisions or political outcomes could not be achieved without fighting, but that such outcomes were ultimately based on the perceived threat or expected effects of physical combat. Therefore, his thinking is appropriate to forms of war such as deterrence or coercion where force is simply threatened; results can be produced 'by the mere possibility of an engagement; the possibility has acquired reality'.[48] This point is made clear when he notes that,

> Combat is the only effective force in war … That holds good even if no actual fighting occurs, because the outcome rests on the assumption that if it came to fighting, the enemy would be destroyed … All action is undertaken in the belief that if the ultimate test of arms should actually occur, the outcome would be favourable. The decision by arms is … in war what cash payment is in commerce … regardless how rarely settlements actually occur, they can never be entirely absent.[49]

Fighting, however, can only be understood as one part of the totality of war, even if it ultimately defines the subject. There are many ancillary activities that either facilitate, are preparatory to or support the act of fighting.[50] However, the idea of fighting has to be a universal feature whether latent or actual, otherwise the whole subject begins to lose its defining character: 'Fighting is the central military act; all other activities merely support it'.[51] So, terrorism and insurgency, while often overwhelmingly political in character, are distinguished from other forms of political resistance by the violent nature of the means employed. The use of force may be sporadic and limited, but it is the threat or use of such violence, employed towards some political object, that lends such conflicts their particular form and makes them comprehensible within the theoretical framework of the trinity: fighting is the 'strand that runs through the entire web of military activity and really holds it together'.[52]

Finally, to stress the centrality of fighting also draws attention to war's material dimension. While it is possible to conceive of groups of humans confronting each other with nothing more than their natural attributes, like a wrestling match writ large, the fact is that from earliest times to the present day, humans have equipped themselves with all kinds of hardware, from weapons and armour to

47 For an excellent illustration of this see, Clausewitz, *On War*, III, 1, p. 213.
48 *On War*, III, 1, p. 212.
49 *On War*, I, 2, p. 111.
50 *On War*, II, 1, p. 146.
51 *On War*, IV, 3, p. 268.
52 *On War*, I, 2, p. 110.

specialised transports and logistical devices. Moreover, such things have to be paid for and troops have to be moved, fed and resupplied. Again, where force is merely threatened this does not negate the influence – in terms of political and psychological effect – of war's tangible, material elements.

The status of the trinity today – a war of ideas

Colin Gray has written, 'I believe neither that Clausewitz wrote the last words that are needed for a fully satisfactory theory of war, nor that some of his analysis could not stand some improvement'.[53] In fact, there is evidence that Clausewitz himself would sympathise with this statement – as Peter Paret notes, he sought to develop a theory that was 'capable of growth, of constant accretion'.[54] His theory, in common with the German philosophical tradition, gained its strength from 'its capacity for further development'.[55] Bassford has noted that 'Clausewitz represents not an end point for speculation about war and its history but a solid foundation for further investigation'.[56]

Most commentators, whether critical or sympathetic, broadly agree that 'few writers have grappled with the sort of fundamental issues so astutely dissected by the great Prussian'[57] or that, 'although the international body of literature on defence studies is vast, general theories are distinguished by their extreme rarity. There are not even many incompetent imitators of Clausewitz'.[58] While many have attempted to breach the Clausewitzian edifice from any number of angles, few have launched an all out offensive in order to impose their own distinctive general theory. This book seeks neither to postulate an alternative to the trinitarian conception, nor argue that Clausewitz 'got it right'. Rather, the central hypothesis lies somewhere in between these two poles, albeit with a strong bias toward the latter.

On the one hand, Clausewitz has endowed us with an unsurpassed theory which we would abandon only to our loss. Admittedly, praise must not obscure the fact that he was writing in a context significantly different and in a time far removed from our own. It would be expecting a great deal of his theory for it to carry the burden of two centuries of turbulent history and rapid change without weakening at the knees somewhat. Yet, as Michael Howard notes, 'It remains the measure of

53 Colin Gray, 'Clausewitz, History, and the Future Strategic World', *The Occasional*, No. 47, June 2003, p. 2.

54 Paret, *Clausewitz*, p. 155.

55 Rudolph Caemmerer quoted in Gat, *Military Thought*, p. 334.

56 Christopher Bassford, 'John Keegan and the Grand Tradition of Trashing Clausewitz', *War in History*, Vol. 1, No. 3, November 1994, pp. 319–36.

57 Steven Metz, 'A Wake for Clausewitz: Toward a Philosophy of 21st Century Warfare', *Parameters*, Winter 1994–95, p. 126.

58 Gray, 'Strategic World', p. 3.

his genius that, although the age for which he wrote is long since past, he can still provide so many insights relevant to a generation, the nature of whose problems he could not possibly have foreseen'.[59] In this respect, Clausewitz is to war as Charles Darwin is to evolution or Adam Smith to economics. All were pioneers and visionaries whose insights remain the standard against which subsequent theories are judged and, despite great strides in methodological techniques and the proliferation of detailed analysis in sub-fields, no holistic theories have successfully usurped their subject defining status.

However, we should not cling dogmatically to any theory that can no longer sufficiently explain its subject in the light of rapid and significant change. Furthermore, in war, resistance to alternative thinking resulting from a dogged belief in one perspective may ultimately cost human lives or even the security of entire communities. Such a view is expressed by Stephen Metz who writes: 'Like adoration for some family elder, the veneration heaped on Clausewitz seems to grow even as his power to explain the world declines'. He claims that 'it is time to hold a wake so that strategists can pay their respects to Clausewitz and then move on'.[60] In this respect, commentators emphasising profound changes in the global social, political and economic environment, and the concurrent transformative impact on the nature of war, are perhaps crucial in keeping us alive – literally and figuratively – to the possibility that, put simply, 'things change'.

So, in recent years, an evident schism has emerged in the field between what may very broadly be termed, 'Clausewitzians' and 'anti-Clausewitzians'. Ironically, the study of war has been debated in a manner resembling the battlefield. The specific reasons given for anti-Clausewitzian arguments are numerous and varied. Some claim that we have entered into an era of so-called 'new wars' in which, due to the impact of globalisation and the rise of identity politics, war has undergone a fundamental change in its nature[61] – war, for instance, is fought for economic gain by actors who wish to sustain the conflict rather than seek victory.[62] Others claim that the rise of non-state warfare, in the form of irregular conflict, insurgency and terrorism renders obsolete a theory formulated in an age of major state warfare.[63] From a different direction, some commentators argue that great changes in war's material aspects occasions Clausewitz's irrelevance as a 'Revolution in Military Affairs' promises to dissipate his famous 'fog of war'.[64]

---

59 Michael Howard, 'The Influence of Clausewitz', in *On War*, p. 49.

60 Metz, 'Wake for Clausewitz', p. 126.

61 See Mary Kaldor, *New and Old Wars: Organized Violence in a Global Era* (Cambridge: Polity, 2002).

62 David Keen, *The Economic Functions of Violence in Civil Wars*, Adelphi Paper 320 (Oxford: Oxford University Press, 1998), p. 11.

63 Martin van Creveld, *The Transformation of War* (New York: Free Press, 1991).

64 See Owens, Bill, *Lifting the Fog of War* (Baltimore: Johns Hopkins University Press, 2001).

Some anti-Clausewitzians, such as John Keegan, Martin van Creveld, Ken Booth and Jan Willem Honig, believe – albeit for different reasons – that the continued adherence to Clausewitzian notions of war is cognitively blinkering, strategically harmful or even poses a threat to civilisation.[65] Honig has argued that Clausewitz stands at the root of a putative modern strategic crisis.[66] All these critiques are united in their belief that, for whatever reason, Clausewitz is no longer, or never really was, a reliable guide for understanding war.[67]

Devotees of Clausewitz generally deny either the depth or extent of the changes recognised by anti-Clausewitzian commentators – hence the claimed continued relevance of *On War* – or believe that critics base their arguments on unfounded conceptions of his theory. Nevertheless, even those who claim continued contemporary relevance for Clausewitz have adopted mistaken interpretations of his trinity. These thinkers are potentially more harmful to our understanding than anti-Clausewitzians. Recently, in an otherwise excellent and timely book by General Sir Rupert Smith, *The Utility of Force*, a similarly misleading presentation of the trinity (as simply people, army, government) runs throughout the work.[68] Regardless of the book's deserved popularity, it raises the prospect of new generations of students taking away partial understandings. This is not to say such interpretations of *On War* are completely wide of the mark – just as with many anti-Clausewitzian writers, they do get some, even many things right, and it is clear that parts of *On War* ostensibly support such representations. Crucially, however, the true worth of his theory only becomes fully apparent through an accurate conception of the whole in all its complexity.

A road to reconciliation?

So, commentators feel increasingly compelled to situate themselves in either one or the other camp and this has tended to generate overstated representations of aspects of Clausewitz's thought. This study suggests there is perhaps room for something of a theoretical reconciliation or at least a narrowing of the apparent scholarly schism. Any putative reconciliation rests on the belief that studies on either side of the divide contain a great deal of value and that, taken together, we

65 Ken Booth, *Strategy and Ethnocentrism* (London: Croom Helm, 1979), Chapter 4; Creveld, *Transformation*; John Keegan, *A History of Warfare* (London: Pimlico, 2004).

66 Jan Willem Honig, 'Strategy in a Post-Clausewitzian Setting', in Gert de Nooy (ed.), *The Clausewitzian Dictum and the Future of Western Military Strategy* (The Hague: Kluwer Law International, 1997), p. 109.

67 A particular irony encountered in anti-Clausewitzian works is that they often – after rejecting Clausewitz – then unwittingly make arguments that are fundamentally Clausewitzian.

68 Rupert Smith, *The Utility of Force: The Art of War in the Modern World* (London: Penguin, 2005).

can perhaps progress towards a clearer understanding of the nature of war. The two camps appear to talk past one another, when in fact there is much room for mutually advantageous dialogue.[69]

On the one hand, most writers who set themselves up as anti-Clausewitzian often do so based on limited and mistaken interpretations (often based on popular secondary analyses which are themselves mistaken).[70] Their studies could benefit from greater historical perspective and a more thorough understanding of the theorist they grandly proclaim to be *passé*. On the other hand, Clausewitzians have perhaps been too resistant to the magnitude and scale of change in war since Clausewitz wrote or too unwilling to consider alternative ideas. Furthermore, the insights of anti-Clausewitzians in terms of observable substantive changes in the *character* of contemporary war are generally entirely valid, even if the precise shape of such change is hotly contested.

At the root of this problem lies our conception of the trinity as this often serves as the central platform for both perspectives. Indeed, as Colin Fleming notes, 'the sense that the nature of military conflict has changed stems directly from debate about the contemporary role of the Trinity'.[71] It appears that the only potentially constructive way forward is to return to the trinity itself and consider the arguments of the two camps based on a more rigorous understanding of its deeper meaning. A theoretical framework based on a detailed interpretation of the trinity is capable of accommodating aspects of war that have emerged as a result of transformations in war's multiple contexts. This book does not engage in detailed empirical analysis of contemporary conflict: existing scholarship is replete with erudite explorations of modern war and strategy. Nor does it seek to directly refute or disprove specific claims of the new wars scholars. While reference will be made to recent studies and debates in the field, the central aim is to provide an accurate interpretation of Clausewitz's theory; in doing so, it should become clear why claims of novelty are perhaps too hasty.

To achieve this, the study focuses on overcoming a number of central barriers to accurate comprehension of the trinity, such as its ostensibly simplistic and misleading presentation, sometimes abstruse or dated terminology and complex substantive content. These areas all require clarification based on a firm grasp of the methodological underpinnings, underlying foundations and assumptions of the trinity. Also, the structure of the whole framework needs to be elucidated, both with respect to its three primary elements and the other levels of war it encompasses; namely war's manifestation in real-world subjects and its multiple contexts. A number of excellent studies have been written along these lines, but they have

69 This proposition is similarly advocated by Colin M. Fleming in his 'New or Old Wars? Debating a Clausewitzian Future', *The Journal of Strategic Studies*, Vol. 32, No. 2, April 2009, p. 238.

70 Fleming argues that few new wars theorists can demonstrate an appreciation of the trinity. Fleming, 'Debating', p. 231.

71 Fleming, 'Debating', p. 217.

generally only been of article or book chapter length.[72] These works provide an excellent foundation, but a deeper and more extensive analysis is required.

The approach here will be underpinned by a close reading of Clausewitz's work guided by three key principles. First, the concepts that comprise the trinity embody a whole range of insights which belies their seeming simplicity. They carry the weight of hundreds of pages of analysis – in *On War* and Clausewitz's other works – and a lifetime of study. The trinity can only truly be understood from this holistic perspective. Second, the analysis adheres to the belief that ideas are inevitably formed as a result of the various experiences of and dominant influences on the thinker in question; they can only be truly comprehended based on a firm knowledge of such factors.[73] Third, our understanding of the trinity will be strengthened by considering its central claims in the light of critical perspectives. The study will refer primarily to a number of prominent authors who challenge his ideas. As the great nineteenth century essayist Thomas Carlyle had it, 'The greatest of faults, I should say, is to be conscious of none'.[74] We must listen to constructive critique attentively and not view *On War* as 'Holy Writ'.[75]

This book also suggests that, even if we reject certain substantive elements of Clausewitz's thought, his *approach* to the subject can serve as a shining example to any student of war. His commitment to historical study, the importance he placed on developing accurate and clear concepts, his struggle to walk the tight-rope of the particular and the universal and, above all, his high standard for ensuring that all propositions in theory must closely reflect the reality of war, all contributed to the inherent worth of his work and stand as testament to his status as a deeply profound thinker.

*

In an unfinished note written in 1827, towards the end of his life, Clausewitz drafted the prophetic statement that, 'If an early death should terminate my work, what I have written so far would, of course, only deserve to be called a shapeless mass of ideas. Being liable to endless misinterpretation it would be the target of much half-baked criticism'.[76] This fact constitutes one of the chief obstacles impeding Clausewitz's modern appreciation, while simultaneously explaining his 'reputation as a profound thinker' because, due to the difficulty of discovering his

72 The best are: Bassford, 'Tip-Toe'; Bassford and Villacres, 'Reclaiming'; and, Gray, 'Strategic World'.

73 Gat, *Military Thought*, p. 256.

74 Thomas Carlyle, 'The Hero as Prophet', in Heroes and Hero Worship (1928). Available online at: <http://www.online-literature.com/thomas-carlyle/heroes-and-hero-worship/2/>, retrieved 12 May 2008.

75 Gray, 'Strategic World', p. 24.

76 Clausewitz, 'Note of 10 July 1827', in *On War*, p. 77.

true meaning, he 'could never have been wrong or less than profound'.[77] This book rests on the assumption that problems of analysis not only derive from the fact that Clausewitz's work is unfinished and intellectually hard to master, but simply because the subject – war – is so vast, endlessly complex and can be approached from so many possible angles. As Clausewitz states, those who realise 'how many vitally important matters are involved in war will understand what unusual mental gifts are needed to keep the whole picture steadily in mind'.[78] Nevertheless, Clausewitz has provided students of war with an invaluable platform from which to commence a journey into the heart of the phenomenon.

77 Gat, *Military Thought*, pp. 254–5.

78 *On War*, III, 1, p. 208.

Chapter 2
Theoretical Foundations[1]

Before proceeding to a detailed analysis of the trinity, it is first necessary to explore the methodological and theoretical foundations on which Clausewitz constructs his concept of the trinity and to determine where his work stands in relation to the broader historical development of military thought. After presenting the central concepts of the trinity, Clausewitz explains that to ensure a dynamic balance between the tendencies, an appropriate approach to the subject is required.[2] As will become clear, the methods he employs to fulfil his intellectual objectives significantly shape the outcome of his work and an appreciation of these foundations is vital for understanding the trinity.

This chapter is divided into two main sections. The first provides a general background to the intellectual environment in which Clausewitz worked and from which he drew many of his methodological devices. It assesses the influences of the Enlightenment spirit of rational and scientific enquiry, the German Movement and the historicist school. The second section examines Clausewitz's approach to theory in greater detail. It considers Clausewitz's views on the purpose and limits of theory and the various methods he employed. It examines in some detail a number of the central theoretical problems he attempted to resolve in his work. The chapter concludes by relating these issues back to the trinity.

Influences

Clausewitz's life straddled a period during which the 'framework of European society … was shaken to its foundations'.[3] He was born in the old regime and laid to rest in the modern world. He not only lived through such turbulent times, but was an active participant in some of the foremost events of the period. As Paret suggests, it is difficult to imagine any age lending itself less to intellectual expressions of

1 This chapter uses material from: Thomas Waldman, 'Clausewitz and the Study of War', *Defence Studies*, Vol. 12, No. 3, June 2012, pp. 345–74, reprinted by permission of Taylor and Francis Ltd (http://www.tandf.co.uk/journals).

2 Carl von Clausewitz, *On War*, trans. and ed. Michael Howard and Peter Paret (New York: Everyman's Library, 1993), I, 1, § 28, p. 101.

3 Michael Howard, *War in European History* (Oxford: Oxford University Press, 1976), p. 75.

universal value.[4] It was easy for those living through such epoch-changing times to either become overwhelmed by the particularities of the age or claim history had reached its grand conclusion. Momentous change and cataclysmic events often induce declarations as to the unprecedented nature of the present at the expense of historical perspective. Indeed, there is strong evidence to suggest that much of the content of *On War* is primarily concerned with Napoleonic warfare and that it was only in his later years that Clausewitz developed a more comprehensive theory. Yet, although Clausewitz's work is inevitably influenced by the character of contemporary warfare, his profound sense of history, sensitivity to change in human affairs and searching mind loosened the tight grip of the present on his thought. If this was true with regard to the events of his day, Clausewitz should equally be considered as one among a number of pioneering military thinkers who were beginning to transcend the intellectual hegemony of the Enlightenment.

Enlightenment: the spirit of enquiry

In general terms, late eighteenth- and early nineteenth-century Germany was a place of profound cultural activity, comparable even to the classical 'golden age' of Athens in terms of the sheer quantity of rich artistic, philosophical and literary output.[5] Such intense activity was largely a reaction to the cultural hegemony of the French Enlightenment and, subsequently, to the upheavals engendered by the French Revolution. Clausewitz was to be heavily influenced by this great cultural backlash, yet our exploration into the roots of his approach must begin in the European Enlightenment.

Although the 'Age of Reason' should be understood as consisting of a confused 'bundle of attitudes' – there certainly was no 'united army of the enlightened'[6] – we can identify some of its major threads. It was, for Kant, the period in which 'Mankind grew out of its self-inflicted immaturity'.[7] Emerging from the scientific and philosophical revolutions of the seventeenth century, it can be seen as a reaction to the superstition, intolerance and religious fanaticism of the medieval world-view. It sought to throw the light of reason on all manner of phenomena through the processes of free thought: the route to truth no longer began with 'Authority', but with doubt.[8]

---

4 Peter Paret, *Clausewitz and the State: The Man, His Theories, and His Times* (Princeton: Princeton University Press, 1985), p. 8.

5 Karl Ameriks (ed.), The Cambridge Companion to German Idealism (Cambridge: Cambridge Unveristy Press, 2006), p. i.

6 John M. Roberts, *The Hutchinson History of the World* (London: Hutchinson, 1976), p. 710.

7 Immanuel Kant, 'An Answer to the Question: What is Enlightenment?', in Immanuel Kant, *Political Writings*, ed. Hans Reiss (Cambridge: Cambridge University Press, 1993), p. 54.

8 Roberts, *History*, p. 710.

At its heart was the optimistic belief in limitless progress through rational processes, the accumulation of knowledge and the assumption that the 'rules' of the physical, human and social worlds could be uncovered through the application of scientific enquiry. Indeed, in all disciplines, an enthusiasm to understand everything and reveal universal truths – typified by the fervent work of the Encyclopaedists – permeated the intellectual environment. 'Dare to know'[9] implored Kant and, as Diderot exclaimed in the *Encyclopédie*, 'Everything must be shaken up, without exception and without circumspection'.[10] Importantly, as Gat states, 'All spheres of human culture and all natural phenomena were to be subjugated to intellectual domination, *and war was no exception*'.[11] The military thinkers of the Enlightenment strove to produce definitive works that would regulate the conduct of military affairs and, as Clausewitz observed, 'Efforts were therefore made to equip the conduct of war with principles, rules, and even systems'.[12] Five thinkers stand out as exemplars of the military-scientific tradition, all of whom Clausewitz was familiar with:[13] Saxe, Guibert, Lloyd, Bülow and Jomini.

Maurice de Saxe opened his 1732 *Reveries de L'Art de la Guerre* with a self-challenge: 'All sciences have principles and rules. War has none'.[14] He believed that war, 'like nature whose secrets had just been laid bare by Newton, operates in accordance with fixed regularities',[15] but that those regularities were, in his time, 'replete with shadows in whose obscurity one cannot move with an assured step'.[16] His work purported to unveil those regularities in the form of principles which, if followed, would lead to victory. Likewise, the French officer and *littérateur* Jacques Antoine Hippolyte, Comte de Guibert, published his influential *Essai tactique générale* in 1772 which aimed to be a grand system of war covering the whole subject.[17] In his *Essai*, he argued that most principles formerly developed were rendered obsolete by changed circumstances and the task of his 'science of

9 Christopher Clark, *The Iron Kingdom: The Rise and Fall of Prussia 1600–1945* (London: Penguin, 2007), p. 247.

10 Tim Blanning, *The Pursuit of Glory: Europe 1648–1815* (London: Penguin, 2007), p. 487.

11 Azar Gat, *A History of Military Thought: From the Enlightenment to the Cold War* (Oxford: Oxford University Press, 2001), p. 30. Emphasis added.

12 *On War*, II, 2, p. 154.

13 Gat, *Military Thought*, p. 174.

14 Maurice de Saxe, *Reveries on the Art of War*, trans. and ed. Thomas R. Phillips (New York: Dover Publications, 2007), p. 17.

15 Martin van Creveld, 'The Eternal Clausewitz', in Michael I. Handel (ed.), *Clausewitz and Modern Strategy* (Abingdon: Routledge, 2005), p. 40.

16 Saxe, *Reveries*, p. 17.

17 R. R. Palmer, 'Frederick the Great, Guibert, Bülow: From Dynastic to National War', in Peter Paret (ed.), *Makers of Modern Strategy: From Machiavelli to the Nuclear Age* (Oxford: Oxford University Press, 1986), p. 106. Martin van Creveld, *The Art of War: War and Military Thought* (New York: Smithsonian Books, 2005), p. 91.

all times'[18] was to apply rigorous methodology to tactics so that, like mathematical truths, people from opposite ends of the world would have to agree on the principles so formulated.[19] Henry Lloyd, meanwhile, wrote in 1781 that 'this art, like all others, is founded on certain and fixed principles, which are by their nature invariable'.[20] He believed mathematical principles could be applied to fortification, artillery, marching and battle formations.[21]

Embracing formalism even further, Heinrich von Bülow's *Spirit of the Modern System of War*, published in 1799, was based on geometrical and mathematical principles concerning an army's relation to its base, its objective and the 'lines of operations' connecting them.[22] As Creveld notes, his work resembles 'nothing so much as a textbook in Euclidean geometry … the entire art of strategy was reduced to a single, simple, geometrical formula'[23] and Howard neatly describes it as 'rococo absurdity'.[24] Clausewitz derided Bülow's 'geometrical system, to which – as all charlatans are wont to do – he ultimately gave a veneer of mathematical elegance'.[25] Antoine-Henri de Jomini was also concerned with discovering a system for the art of war, which, like Bülow, relied heavily on concepts such as 'base' and 'lines of operations'.[26] In particular, Jomini drew attention to the importance of operating on so-called 'interior lines'.[27] In contrast to earlier thinkers, Jomini insisted that tactical matters were subject to considerable flux, while strategy 'may indeed be regulated by fixed laws resembling those of the positive sciences'.[28] He sought his principles of war by 'abstracting it from its political and social context … turning warfare into a huge game of chess'.[29] In many respects, Jomini represented the culmination of the formalistic and schematic approach; he was almost evangelical in his insistence on the 'timeless verities' that his science of war would provide.[30]

It would be unfair to claim that this was all these theorists contributed to the study of war or that Clausewitz was simply reacting against, rather than building

18 Guibert quoted in Palmer, 'National War', p. 107.

19 Gat, *Military Thought*, p. 49.

20 Henry Lloyd, *History of the Late War between the King of Prussia and the Empress of Germany and her Allies* (London: 1781), p. vi.

21 Gat, *Military Thought*, p. 72.

22 Palmer, 'National War', p. 115.

23 Creveld, *Military Thought*, p. 97.

24 Michael Howard, *Studies in War and Peace* (London: Temple Smith, 1970), p. 25.

25 Carl von Clausewitz, 'On the Life and Character of Scharnhorst', in Clausewitz, *Writings*, p. 103.

26 Brian Bond, *The Pursuit of Victory: From Napoleon to Saddam Hussein* (Oxford: Oxford University Press, 2006), p. 44.

27 See Antoine Henri de Jomini, *The Art of War*, trans. G. H. Mendell and W. P. Craighill (Texas: El Paso Norte Press, 2005), pp. 79–100.

28 Jomini, *Art of War*, p. 258.

29 John Shy, 'Jomini', in Paret (ed.), *Makers*, p. 144.

30 Shy, 'Jomini', p. 145.

on, their ideas. In many respects, Clausewitz's most penetrating insights emerged precisely because existing conceptions provoked him to articulate why and how they were mistaken. These earlier works certainly are more nuanced than outlined above and we should avoid subscribing to a 'vague and stereotyped image of Clausewitz's predecessors'.[31] For instance, Jomini was aware that there was more to war than his formalistic concepts conveyed. He even admitted that 'everything that can be termed the poetry and metaphysics of war will have a permanent influence on its results' and that 'theories cannot teach men with mathematical precision what they should do in every possible case'.[32] Yet, although Jomini acknowledged that 'war, far from being an exact science, is a terrible and impassioned drama', he ultimately believed it was regulated 'by three of four general principles'.[33] The implication is that, as Sun Tzu put it, 'Those who know [the principles] prevail, those who do not know them do not prevail'.[34] In the hands of competent generals with good troops they were said to constitute 'means of almost certain success'.[35]

So, regardless of any qualifications, the overarching feature of these theories was their emphasis on uncovering quantifiable, formal principles. Their recognition of other less tangible factors did not greatly alter the regular operation of the principles they had purportedly discovered; moral, psychological or political factors were not integrated into their theories, but presented as addendums, secondary concerns or awkward dimensions that theory might ignore. Clausewitz rejected such 'fanciful', 'one-sided' and 'pretentious' theories, accusing them of 'sham brilliance and sterile pedantry'.[36] Furthermore, those 'narrow systems' were accompanied by a 'lawless rabble of camp followers' in the form of '*jargon, technicalities, and metaphors*'.[37] Their weakness, according to Clausewitz, was a consequence not only of their search for unattainable scientific principles, but also their inevitable *selectivity*: their objective could only be attained by arbitrarily restricting their inquiries to those phenomena that would yield to precise formulae – or what T. E. Lawrence would later term war's 'algebraical elements'.[38] Clausewitz, by contrast, although certainly recognising the 'mathematical' elements of strategy,[39] emphasised the importance of understanding the whole and all the 'endless complexities involved'.[40] Moreover, Clausewitz recognised such systems were beset by a fundamental failing whereby, in Brian Bond's words, 'if

---

31 Gat, *Military Thought*, p. 95.

32 Jomini, *Art of War*, p. 258 and p. 260.

33 Jomini, *Art of War*, p. 286.

34 Sun Tzu, *The Art of War*, trans. T. Cleary (Boston: Shambhala, 1988), p. 45.

35 Jomini, *Art of War*, p. 260.

36 Clausewitz, 'On the Life and Character of Scharnhorst', pp. 103–4.

37 *On War*, II, 5, p. 197.

38 Thomas E. Lawrence, 'The Evolution of a Revolt', *Army Quarterly and Defence Journal*, October, 1920.

39 *On War*, III, 2, p. 215.

40 *On War*, II, 2, p. 154.

strategy was a science whose principles could be learnt what was to prevent all the belligerents learning them?'[41]

Yet, as John Keegan states, Clausewitz should be seen as 'a child of the Enlightenment' in that he grew up in a Europe still heavily influenced by its message, despite the fact that it had passed its zenith by the latter quarter of the eighteenth century.[42] It must also be remembered that Germany experienced it own unique version of the European Enlightenment: the *Aufklärung*. As Tim Blanning notes, 'It is difficult to comprehend how anyone familiar with the *Aufklärung* can think that it was part of a single Enlightenment'.[43] The German Enlightenment was altogether more realistic, serious and profound. It was also less radical, whilst still being preoccupied with shining the light of reason on all manner of phenomena. Importantly, it was not essential to subscribe to the fullest excesses of the French Enlightenment to be affected by the spirit of reason.

Tellingly, only a thinker influenced by the paradigmatic climate of Enlightenment inquiry would feel compelled to write of his work that 'its scientific character consists in an attempt to investigate the essence of the phenomena of war … No logical conclusion has been avoided'.[44] Indeed, Strachan notes that *On War* represents the culmination of the Enlightenment's influence on military thought.[45] Certainly Clausewitz was too wise to underrate the rich vein of thought that emerged from the 'classical' theorists.[46] Equally, Clausewitz did not fully subscribe to the views of the so-called 'antinomians', led by the arch-sceptic Berenhorst, who held that formal theory was inapplicable to war. Rejecting such defeatism, Clausewitz was prone to abstract speculation and motivated by an 'intense search for "scientifically valid" methods of interpreting the reality of war'.[47] This was evident in his early work, such as his 1805 article criticising Bülow, in which he states that, 'We shall apply ourselves to general notions … because in the constitution of a science it is to them that we shall above all return'.[48]

Strachan captures this aspect of Clausewitz's approach when he notes that 'he had read the works of the Enlightenment, and, for all his damning comments about certain military theorists, he was determined to develop a theory of his own'.[49] We should perhaps view Clausewitz as the illegitimate heir of a tradition of military

41 Bond, *Pursuit*, p. 47.

42 John Keegan, *A History of Warfare* (London: Pimlico, 2004), p. 12.

43 Blanning, *Pursuit of Glory*, p. 499.

44 Clausewitz, 'Author's Preface To an Unpublished Manuscript on the Theory of War, Written between 1816 and 1818', in *On War*, p. 69.

45 Hew Strachan, *Carl von Clausewitz's On War: A Biography* (New York: Atlantic Monthly Press, 2007), p. 89.

46 Howard, *Studies*, pp. 26–7.

47 Paret, *Clausewitz*, p. 148.

48 Quoted in Raymond Aron, *Peace and War* (New Jersey: Transaction, 2003), p. 43.

49 Hew Strachan, 'Clausewitz and the Dialectics of War', in Hew Strachan and Andreas Herberg-Rothe, *Clausewitz in the Twenty-First Century* (Oxford: Oxford University Press, 2007), p. 19.

thought stemming from the French Enlightenment, via the German *Aufklärers* of the 1770s and 1780s, and through to his contemporaries, including prominent figures such as Jomini and his influential mentor, Scharnhorst. This was a lineage united by the search for a general theory of war but divided by their contrasting and individualistic approaches. As Clausewitz reminds us, he was reacting against what he believed to be the mistaken views of excessively rationalist theorists, yet importantly, and in a more positive sense, they certainly provided him with a guiding vision of intellectual clarity, systematic analysis and objective inquiry, without which *On War* may never have been conceived.

Here we must cease discussion of Clausewitz's indebtedness to the Enlightenment lest we overstate the association. Clausewitz owed more to a tradition that represented, in many respects, its very antithesis. Following earlier thinkers, Clausewitz certainly developed a 'system', but it was more a 'system of notions and propositions which are linked and which translates into ideas the structure of the object'.[50] As Paret notes, he 'rejected the popular Enlightenment, with its doctrinaire faith in rationality and progress, and found no difficulty in acknowledging limits to human understanding'.[51] To reduce war to a list of rules was mistaken: no formalistic system could capture the incredible complexity of the phenomenon. For Strachan, it was perhaps less the influence of any 'new thinking', but rather Clausewitz's own experiences of war and his growing familiarity with military history that constituted the 'reality checks on [his] inclination to abstraction'.[52]

An alternative worldview: the German Movement

While Strachan is correct, we should nonetheless emphasise the extent to which the relative uniformity of Enlightenment military thought, typified by men such as Bülow and Guibert, was being assailed by thinkers such as Berenhorst and Scharnhorst, thus opening Clausewitz's mind to original ideas, while the pervasive cultural forces of the 'German Movement' were revealing a whole new perspective on reality to him. German society was experiencing a profound if underlying rupturing of established patterns as the prevailing order was beset by various competing socio-cultural forces. Clausewitz's work reflects this tension.

In military terms, Berenhorst was an important counterweight to the rationalist hegemony and his contrarian approach would have impressed Clausewitz's open and searching mind. Berenhorst, in his *Reflections on the Art of War*, published between 1796 and 1799, argued that there could be no fixed, scientific principles in war and suggested instead that everything was dependent on the operation of

50 Raymond Aron, *Clausewitz: Philosopher of War*, trans. C. Booker and N. Stone (New Jersey: Prentice-Hall, 1985), p. 55.

51 Paret, *Clausewitz*, p. 149.

52 Strachan, 'Dialectics', p. 19.

individual genius that could be neither imitated nor analysed.[53] The search for immutable laws was wrong-headed in his opinion: 'What is the use of rules when one is covered up to one's ears with exceptions?'[54] War was suffused with the unpredictable and incalculable. Blind chance, the spirit of an army, the moral factors that animate the troops: these 'unknown and uncontrollable modifications of the human spirit' were the true elements of war.[55] In military terms, Berenhorst was the embodiment of the Counter-Enlightenment. If Jomini and Berenhorst typify, respectively, the extremes of rationality and irrationality in war then it was left to Clausewitz to reconcile these opposing strands of thought.[56]

The influence of Scharnhorst was vital, for at a crucial time in the development of Clausewitz's thought, his great friend and mentor was beginning to rail against the one-sided theories that had emerged at around the turn of the century. Scharnhorst presented Clausewitz with the possibility of a theoretical middle-ground whereby war has two sides: 'one is mechanical and susceptible to theoretical study, the other circumstantial and dominated by creative genius and experience'.[57] Thus, we can begin to understand that Clausewitz's approach reflected the scientific Enlightenment standard principally in terms of his foundational methodology and the desire to unearth the universal elements of war. However, within this broad conception, he consistently expressed the *limits of theory* and the reach of rules or law-like propositions. As a result, his work contains some paradoxical features: for instance, his concrete proposition that the 'uncertain' in war must be recognised as central to theory. Clausewitz believed it was dangerous to generalise from ideas that have only limited validity and this assertion constitutes a central theme in his work: the search for an appropriate balance between the universal and the particular. Vital to the resolution of such issues would be a robust understanding of history, in terms of both substance and deeper philosophical meaning.

Rediscovering history

In broad terms, historical perspective had been lacking in Enlightenment thought with its strong bias towards the future, typified by the belief in progress based on the discovery and implementation of rational principles. The past was viewed as the backward place we had come from, the future the 'sunlit uplands' we were heading towards. Indeed, various strains of social utopianism stem from such rationalist, a-historical thinking.[58] With regard to war, such thinking has manifested itself in the belief that it could be perfectly understood, limited or, indeed, eliminated

53 Howard, *Clausewitz*, p. 23.

54 Quoted in Gat, *Military Thought*, p. 156.

55 Gat, *Military Thought*, p. 155.

56 Creveld, *Military Thought*, p. 107.

57 Gat, *Military Thought*, p. 167.

58 See John Gray, *Black Mass: Apocalyptic Religion and the Death of Utopia* (London: Penguin, 2007).

from human social relations. This occurs because, as Jeane Kirkpatrick explains, 'Thought set free from experience is unlimited by the constraints of experience or of probability. If history is not relevant, then the future is free from the past. Therefore, theories cut loose from experience are usually blindly optimistic'.[59]

This is precisely the kind of thinking Clausewitz rejected. The excesses generated by Enlightenment thought had been revealed to him practically through the murderous consequences of the French Revolution and theoretically in the limited theories of rationalist thinkers who conveniently banished uncertainty, political conditions and emotions from their work. The German Movement that emerged in the late nineteenth century signified a dramatic shift away from such a-historical perspectives and 'moralising assumptions'.[60] This attitude toward the past was symptomatic of the period and found its greatest expression in the work of Johann Gottfried von Herder.

Herder was a dominant figure in the development of the conception of history known as historicism. The essential premise of this outlook was its emphasis on the particularity and uniqueness of historical cultures and societies, which could only be truly understood in their own right through empathetic analysis of their peculiar characteristics. This individualising view of the past stood in stark contrast to the prevailing Enlightenment conception, with its deterministic belief in progress and the universalising, almost patronising, projection of its own values back through history. Herder believed that the '*Volksgeist*'[61] or 'material culture of a people's life, their skills, language and beliefs and artistic and literary practices, make up a single self-sufficient and characteristic whole'.[62] Herder's concern with the importance of comprehending historical conditions was imparted to Clausewitz through Scharnhorst and the theological scholarship of Friedrich Schleiermacher.[63]

Arguably the 'greatest influence on the development of Clausewitz's historical and theoretical thought,'[64] Scharnhorst had emphasised that history was central to the study of war.[65] As Clausewitz notes in his obituary, 'On the Life and Character of Scharnhorst', the latter had a 'great preference and respect for the power of historic evidence in all matters that preoccupied him'.[66] Although the style of their historical work differed in both style and content, the example of Scharnhorst is clearly apparent in Clausewitz's pragmatic, objective and non-partisan approach to

59 Kirkpatrick, *Democracy and Double Standards*, p. 10.

60 Paret, *Understanding*, p. 138.

61 Marvin Perry, *An Intellectual History of Modern Europe* (Boston: Houghton Mifflin Company, 1993), p. 234.

62 Nicholas Boyle, *German Literature: A Very Short Introduction* (Oxford: Oxford University Press, 2008), p. 48.

63 Schleiermacher was an acquaintance of Clausewitz.

64 Paret in Clausewitz, *Writings*, p. 85.

65 Paret, *Clausewitz*, p. 71.

66 Clausewitz, 'On the Life and Character of Scharnhorst', p. 100.

historical study. Essentially, Scharnhorst taught Clausewitz that the only evidence we have about war is historical. In *On War*, Clausewitz reaffirms this lesson: 'Undoubtedly, the knowledge which is basic to war is empirical ... revealed to us only by experience'.[67]

Paret notes that Clausewitz had a clear 'sense of the particularity of individuals and societies, which observed their own laws rather than rationalist abstractions'. While he held that human feelings might have a universal nature, 'their human, social, and intellectual expression was in flux'.[68] *On War* is scattered with historical references, short analyses and methodological enquiries into the proper use of history. The extent of his historicism is most impressively revealed in his sociological survey of war from the 'semibarbarous Tartars' through to the nineteenth century in Chapter 3B of Book 8. He concludes that chapter by stating that, 'we wanted to show how every age had its own kind of war, its own limiting conditions, and its own peculiar preconceptions ... it follows then that the events of every age must be judged in the light of its own peculiarities'.[69]

Explorations

The preceding survey of the dominant influences on Clausewitz's approach provides the basis for a more detailed examination of the method that underpins *On War*. What is fascinating about his approach is the way in which it may be understood as a synthesis of the various pre-existing strands of thought. His theoretical work was derivative, but his originality lay in the manner in which he combined distinctive analytic strands and applied their integrated force to issues surrounding war.[70] We might visualise Clausewitz as being tied to a number of ropes pulling him in various directions and, while at times he seemed to be drawn one way or the other, ultimately he managed to retain his balance, control those various influences and utilise their strongest elements to achieve his ultimate purpose. This was no easy task. At many points, the temptation to be drawn toward a particular method was strong, and this tension can be sensed in his work.

The following sections consider some of the central themes that emerge in Clausewitz's methodology and, in doing so, it will largely let Clausewitz speak for himself through the selective use of pertinent quotes from his works. First, his general conception of theory is outlined. Subsequently, a set of five methodological pairs are presented as platforms for examining aspects of Clausewitz's approach: art versus science; dogmatism versus pedagogy; theory versus reality; material versus moral factors; and the universal versus the particular.

---

67 *On War*, II, 6, p. 199.

68 Paret, *Clausewitz*, pp. 166–7.

69 *On War*, VIII, 3B, p. 717.

70 Paret, *Clausewitz*, p. 150.

Developing a conceptual method

Scharnhorst had imparted to Clausewitz the importance of military theory which, through conceptualisation, 'makes possible the intellectual treatment of the factors active in war'.[71] This sentiment is clearly apparent in Clausewitz's work. As he states, 'this subject, like any other that does not surpass man's intellectual capacity, can be elucidated by an inquiring mind, and its internal structure can to some degree be revealed. That alone is enough to turn the concept of theory into reality'.[72] Despite his disdain for the 'systems' of Enlightenment military thinkers, Clausewitz did not follow Berenhorst in his theoretical antinomianism: war, he held, was susceptible to theoretical analysis. The fundamental task of the theorist was to 'eliminate the weeds that spring from ignorance' by shining the light of theory on all phenomena.[73]

According to Clausewitz, the task of theory was to clarify reality or, more accurately, to help men clarify it by stimulating their minds: 'Theory will have fulfilled its main task when it is used to analyse the constituent elements of war, to distinguish precisely what at first sight seems confused'.[74] He also remarked that the 'human mind ... has a universal thirst for clarity, and longs to feel itself part of an orderly scheme of things'.[75] While rejecting the idea that war itself could be treated as a science, this did not mean it could not be *approached* in the spirit of scientific inquiry, only that one could not expect law-like principles to emerge as a result. Clausewitz proposes a systematic and clear-headed approach that would serve to 'clarify concepts and ideas that have become, as it were, confused and entangled'.[76]

An essential element, even precondition, of reliable theory involves an attempt to recognise the defining character of phenomena: he states that, 'Not until terms and concepts have been defined can one hope to make any progress in examining the question clearly and simply'.[77] His theory would seek to develop an understanding of war through a detailed exploration of its constituent elements, as expressed at the very beginning of *On War*: 'I propose to consider first the various elements of the subject, next its various parts or sections, and finally the whole in its internal structure'.[78] This, he explains, enables us to make a 'differentiation and identification' of the 'major components' of the structure of theory.[79] How was he to achieve this goal? It is useful to consider the principal methodological

71 Gat, *Military Thought*, p. 162.
72 *On War*, II, 3, p. 174.
73 Hugh Smith, *On Clausewitz* (Hampshire: Palgrave Macmillan, 2005), p. 176.
74 *On War*, II, 2, p. 163.
75 Clausewitz, 'Unfinished Note', in *On War*, p. 80.
76 *On War*, II, 2, p. 152.
77 *On War*, II, 2, p. 152.
78 *On War*, I, 1, § 1, p. 83.
79 *On War*, I, 1, § 28, p. 101.

tools Clausewitz employed in order to resolve the theoretical problems he faced, because knowledge of this is vital to understand the development of his ideas. This subject has generated a great deal of debate: was Clausewitz Kantian, Hegelian or even Newtonian in his approach? [80] It is important to note that a definitive answer on this issue is probably unrealisable, primarily because Clausewitz did not subscribe *to any one* particular method and never absolutely or in the precise way intended by its originator. What is not in doubt, however, is that certain central methodological tools are apparent: foremost amongst these is the dialectic.

The use of the dialectical method was a strong feature of German Idealist philosophy of the time and is most strongly associated with Hegel. It appears to have been a method that Clausewitz only seriously appropriated later in his life as he attempted to deal with issues that involved tensions between various concepts.[81] Where he employed dialectical arguments to draw out ideas, the aim was to 'achieve understanding through debate, through point and counterpoint; the dialogue is continuous'.[82] Through this line of reasoning, a thesis and its antithesis leads to a logical contradiction and must therefore be resolved in some form of reconciliation.

The influence of the dialectic on his approach – although not in a strictly Hegelian form – is clearly revealed in Book 8 when Clausewitz states that, 'Up to now we have considered the incompatibility between war and every other human interest … that no philosophy can resolve. Now we must seek out the unity into which these contradictory elements combine in real life, which they do by partly neutralising one another'.[83] Notwithstanding the absence of a Hegelian progression to an 'Absolute Idea' and the lack of 'clear dialectical sign posts, labelled thesis, antithesis, and synthesis',[84] failure to understand Clausewitz's method leads to considerable misunderstanding and, in particular, the 'inability of his readers to proceed beyond his thesis to its antithesis and to understand that he usually locates the truth in the tension between the two'.[85] The trinity might be conceived as a form of synthesis in this respect, but it is one that embraces continuing tension between concepts rather than any final resolution.

So, Clausewitz approached the phenomenon of war by breaking it down into observable and definable concepts; a vital process if theory was, as he describes, 'to serve its principal purpose of discriminating between dissimilar elements'.[86] In order to determine the operation of those concepts, Clausewitz often employed

80 Creveld, 'The Eternal Clausewitz', pp. 35–48.

81 Prior to the last stage of his life, there 'are no theses and antitheses, no polarity or dialectic, not, indeed, any mention of "absolute war"'. Gat, *Military Thought*, p. 233.

82 Strachan, 'Dialectics', p. 37.

83 *On War*, VIII, 6B, p. 732.

84 Christopher Bassford, 'The Primacy of Policy and the "Trinity" in Clausewitz's Mature Thought', in Strachan and Herberg-Rothe, *Clausewitz*, p. 75.

85 Paret, *Understanding*, pp. 163–4.

86 *On War*, II, 2, p. 147.

a form of dialectical reasoning, which posed sets of opposites against one another and sought to explore the nature of their relationship. Reflecting this in a broader sense, a number of oppositional pairs can be observed in Clausewitz's methodology, each of which we will explore in turn. Together these five pairs illuminate Clausewitz's endeavour to discover the points at which such apparent tensions cease to contradict each other.

Art versus science

Clausewitz believed the answer to whether war was an art or science was of great significance, even 'indispensable' because one's conclusion determined the way in which the entire subject is approached.[87] This problem of course reflects the broad dichotomy that Clausewitz recognised as existing within military theory. Thinkers such as Bülow sought scientific precision and ended up with limited, one-sided theories. Thinkers such as Berenhorst believed war was akin to a creative art, entirely dependent on individual genius, and ended up belittling the role of theory. If neither of these approaches would suffice, Clausewitz endeavoured to determine why. Book 3, Chapter 3 of *On War* is taken up with a detailed discussion of these issues.

Clausewitz first sought to establish the difference between the two disciplines by examining the prime objectives of each. In science, the object is 'pure knowledge' based on inquiry and is principally associated with subjects that aim at fixed values, the discovery of regularities and the identification of laws of logic or cause and effect. Pure science has no objective other than the furtherance of knowledge. In art, the object is 'creative ability'. Art, Clausewitz held, is intended to produce an aesthetic effect, to which end it is necessary to use given means. To combine them is to create; art is thus the capacity to create. Unlike science, art is not susceptible to law-like formulas, primarily because it aims at effects on the mind. Yet, for him, these basic distinctions are misleading. All arts contain 'discrete sciences', such as the geometrical elements of architecture, whereas applied science always involves an element of creativity and judgement, thus, it 'is impossible to separate art and knowledge altogether'. Nevertheless, the basic distinction between, on the one hand, 'inquiry and knowledge' and, on the other, 'creation and production' is clear.[88] So how does this apply to war?

He explains that, in war, like no other activity, the complete and 'total assimilation' of knowledge and ability is vital. For the commander, whatever thorough knowledge he may have absorbed from so many scientific military manuals, all will be useless unless it can be turned into 'genuine capability'. Books may lay down the basic principles of an art but they cannot teach us 'how to do anything'. War is not even comparable to a mechanical art or craft, whereby given truths can be objectively and impersonally applied to the matter in hand. Continual

87 *On War*, II, 3, p. 173.

88 *On War*, II, 3, pp. 172–4.

change and the need to respond to it compels the commander to carry the whole intellectual apparatus of his knowledge with him'.[89] In war, knowledge and ability must be alloyed. It is never enough simply 'to know' because 'absolute, so-called mathematical, factors never find a firm basis in military calculations'.[90] Likewise, creative ability is dependent on a knowledge of the relationship of things and the likely effects of given actions.

So war is strictly neither a pure art nor a pure science. It will always involve a complex interplay of the two, particularly when approached from the perspective of the individual commander. It appears that, were Clausewitz forced to make a decision, he would argue war will always be more like an art – particularly at its higher levels – primarily because of its practical nature; a conclusion that even Jomini, despite his positivism, arrived at.[91] But Clausewitz was not content to let the matter rest there. In fact, he believed that to base understanding on the distinction between art and science 'is misleading in that it has unintentionally caused war to be put on par with other arts or sciences, resulting in a mass of incorrect analogies'.[92]

Such analogies are misleading because neither discipline captures war's unique nature. Although it may involve elements of both, it is altogether different because it is a '*conflict of living forces*'.[93] Previously, both the art and science of war viewed it as a 'unilateral activity' or as 'nothing but a gradual rise from a craft to a refined mechanical art'.[94] Clausewitz explains why war is, in fact, unique:

> The essential difference is that war is not an exercise of the will directed at inanimate matter ... In war, the will is directed at an animate object that reacts. It must be obvious that the intellectual codification used in the arts and sciences is inappropriate to such an activity.[95]

If war is not usefully understood as science or art, what is it? War, he explains, is 'part of man's social existence'. He suggests a better analogy may be commerce, but even that ignores its fundamental reality: that it is a 'clash of major interests, which is resolved by bloodshed'. War is deeply human and social. It takes place in a climate of danger, fear, exertion and uncertainty and is inherently reactive. Art and science may, to some limited extent, display elements of these characteristics, but taken together war is something altogether distinctive. If it reflects any other

89 *On War*, II, 2, pp. 170–71.

90 *On War*, I, 1, § 21, p. 97.

91 'Strategy, particularly, may indeed be regulated by fixed laws resembling those of the positive sciences, but this is not true of war viewed as a whole'. Jomini, *Art of War*, p. 258.

92 *On War*, II, 3, p. 173.

93 *On War*, II, 3, p. 174. Emphasis added.

94 *On War*, II, 2, p. 153.

95 *On War*, II, 3, pp. 173–4.

human activity it is *politics*, because after all, as he explains, that is 'the womb in which war develops – where its outlines already exist in their hidden rudimentary form, like the characteristics of living creatures in their embryos'.[96]

This presented Clausewitz with a serious problem. Namely, to what extent can theory provide laws regarding an activity involving a dynamic interaction of human forces and so universally pervaded with chance? Clausewitz's consideration of this issue led him to a particular conception regarding the purpose and limits of theory.

Dogmatism versus pedagogy

As Strachan has noted, 'much of *On War* is written with a forthrightness that can easily be mistaken for dogma'[97] and Raymond Aron has observed that, 'It is the systematic character of the Treatise which has sometimes made superficial readers confuse it with dogmatism'.[98] It is clear that 'dogmatic' was precisely the epithet Clausewitz wished to avoid. This is plainly revealed in the heading of a section entitled, 'Theory Should Be Study, Not Doctrine'. In it, he states that 'a theory need not be a positive doctrine, a sort of *manual* for action' but should be seen more as a 'guide'. Far from dictating exactly what the commander should do on the battlefield, it is intended more to 'light his way, ease his progress, train his judgement, and help him to avoid pitfalls'.[99] Clausewitz emphatically states that, 'it is simply not possible to construct a model for the art of war that can serve as a scaffolding on which the commander can rely for support at any time'.[100] History's great commanders did not rely on such guidance, indeed, 'anyone who thought it necessary to begin the education of a future general with a knowledge of all the details has always been scoffed at as a ridiculous pedant'.[101]

Here he differed from his contemporary Jomini who, in the conclusion to his work, stated that, 'Correct theories, founded on right principles ... will form a true school of instruction for generals'.[102] In contrast, Clausewitz was not so interested in telling people *what* to think than in guiding them in *how* to think. This belief has its analogue in the critical analysis of past wars. The critic, Clausewitz insisted, should not study a particular case through a 'mechanical application of theory', but rather understand that the suitability of any theory is a matter of judgement. A critic should never use theory as law, but only – as the soldier does – as *aids to judgement*'.[103] Just as 'scientific guidelines' are inappropriate tools on the

96 *On War*, II, 3, p. 173.
97 Strachan, 'Dialectics', p. 17.
98 Aron, *Clausewitz*, p. 51.
99 *On War*, II, 2, pp. 162–3.
100 *On War*, II, 2, p. 161.
101 *On War*, II, 2, p. 168.
102 Jomini, *Art of War*, p. 261.
103 *On War*, II, 5, p. 183.

battlefield, so they are in critical analysis where truths are grasped 'through the natural perception of the mind'.[104]

So, it was not Clausewitz's aim to develop a utilitarian theory, but one that was predominantly analytic, descriptive and didactic: theory can teach but it cannot prescribe. What was the point of producing a theory which genius could laugh at and that the man of limited ability could learn by rote but fail to convert into success? For him, 'It is the theory which indicates that the responsibility for decision is thrown back on the man of action'.[105] This was an approach consistent with the contemporary German notion of *Bildung*, or self-direction and autonomous judgement.[106] Theory can offer sound insights into the 'great mass of phenomena' associated with war, but then leaves the mind to 'rise into the higher realms of action'.[107] Theory teaches that the essence of genius is that it is creative, not imitative; it can transcend existing rules, but in doing so it creates new ones. Furthermore, Brodie contrasts Clausewitz's approach with that of the modern army field manuals which seek to capture centuries of experience in a 'few tersely worded and usually numbered "principles of war"', concluding that 'Clausewitz would have been appalled at such attempts, and not surprised at some of the terrible blunders that have been made in the name of those principles'.[108] As General Ulysses S. Grant once commented, 'If men make war in slavish obedience to rules, they will fail'.[109]

Clausewitz did not set down strict rules that should be followed unconsciously or without regard to circumstances. Book 2 of *On War* almost amounts to a plea not to read his work in that way. Commanders, faced with the problems of the moment, will always search for direct routes to success and are wont to skate over sections that do not provide such practical and, indeed, psychological comfort. And here is the irony. The very weakness of the intellectual capacity of commanders not imbued with natural talent would be precisely those desperate to read their way out of difficult situations, looking for the type of principles that Clausewitz rejected. He would not provide such petty rules to follow because 'petty things will make a petty mind' and a petty mind was deemed insufficient for a commander faced with the complexity of war.[110]

---

104 *On War*, II, 5, p. 196.

105 Aron, *Clausewitz*, p. 55.

106 Terry Pinkard, *German Philosophy 1760–1860: The Legacy of Idealism* (Cambridge: Cambridge University Press, 2008), pp. 7–9.

107 *On War*, VIII, 8, p. 698.

108 Bernard Brodie, *War and Politics* (London: Cassell & Company Ltd, 1974), p. 446.

109 Quoted in Hew Strachan, *European Armies and the Conduct of War* (Kings Lynn: Routledge, 2006), p. 3.

110 *On War*, II, 2, p. 168.

This is not to deny that Clausewitz does indeed articulate various principles or that the weight of his ideas lean towards favouring a certain form of warfare.[111] However, principles in *On War* always remain conditional. The limits to his 'system' are repeatedly expressed, often through metaphors that should not escape even the casual readers eye. Moreover, in almost all cases where he expresses one principle or another, no matter how forcibly stated, they are accompanied by qualifications. As Beyerchen notes, even his 'most Newtonian-sounding analogy of a "centre of gravity" becomes swamped in qualifications and caveats intended to convey the complexity of war'.[112]

So, Clausewitz did not aim at a positive theory of war, nor did he seek to provide a crutch for petty-minded leaders. For him, laws were not applicable in war, as they may be in the sciences. He emphasised the difficulty of developing concrete theory because of the often large distance between cause and effect, the continuous interaction of opposites, the impact of changing conditions and the play of intangible forces that theory could never quite define.[113] Yet, despite these limitations, as Creveld notes, theory 'can aspire to save the strategist from the need to think out everything from the beginning, and provides him with a starting point for thought'.[114] Importantly, Clausewitz's anti-doctrinaire view depended on his contention that it should not conflict with the reality of war.

Theory versus reality

In his panegyric to his late friend and mentor, Clausewitz described the qualities of Scharnhorst's mind which made for such 'apt and forceful ideas':

> At the point where imagination leads the brilliant systematizer beyond specifics, Scharnhorst would quietly turn back and direct his energies toward reconciling ideas and reality, carefully fusing the two by theoretical and historical analysis … In war mere imagination has no creative power at all ... [115]

This quote lucidly reflects the approach Clausewitz ultimately adopted himself, while simultaneously revealing his huge debt to Scharnhorst.[116] In criticising flights of the imagination, Clausewitz perhaps had in mind the German Idealists. Although not unsympathetic towards their general project, he was adamant than in

111 Clausewitz states that a 'whole range of propositions can be demonstrated without difficulty'. Clausewitz, 'Unfinished Note', p. 80.

112 Alan Beyerchen, 'Clausewitz, Nonlinearity and the Unpredictability of War', *International Security*, Vol. 17, No. 3, Winter, 1992, p. 84.

113 *On War*, I, 7, p. 140.

114 Martin van Creveld, *The Transformation of War* (New York: Free Press, 1991), p. 116.

115 Clausewitz, 'On the Life and Character of Scharnhorst', p. 100.

116 Paret, *Clausewitz*, p. 71.

such a practical activity as war, pure abstract theorising would lead the thinker into 'dazzling and flowery fantasies' that would have little applicability in reality.[117]

In an essay entitled 'The Germans and the French', Clausewitz commented how the German intellect, 'instead of taking immediate pleasure in the correspondence of its ideas to reality … delves deeply into the nature of things, into abstractions, and strives for complete understanding'. This, he says, 'often damages his usefulness in practical, particularly, political life'.[118] For Clausewitz, such 'presumptuous philosophy' is worthy of contempt because, not only is it so completely detached from the real and pressing concerns of the day, but, moreover, individual generations 'do not exist to observe the world; by constantly striving for rational goals they *are* the world'.[119]

Clausewitz's practical nature convinced him of the futility of speculative notions that conflicted with reality and that lead to a situation where an author 'soothes himself with obscure ideas which would not satisfy him if expressed in plain speech'.[120] Scharnhorst taught his student to 'oppose artificial and learned theorising by encouraging a certain naturalness of thought, which defeats empty phraseology and brings the false conflict between theory and practice to an end'.[121] What abstract theorising there is in *On War* – and there is plenty, particularly in Books 1 and 8 – is always balanced by a forceful collision with reality.

The way in which Book 1, Chapter 1 sways between the abstract and the concrete can be a cause of mental 'sea-sickness' for some, but that is simply the result of a mind that was never content to rest in the comfort of one or the other, preferring instead the challenge of searching out the points at which the two cease to contradict each other. This is why recognising the dialectic in Clausewitz's approach is so important – without it one cannot understand what he was trying to achieve. In an often quoted passage, Clausewitz held that he never avoided logical conclusions,

> but whenever the thread became too thin I have preferred to break it off and go back to the relevant phenomenon of experience. Just as some plants bear fruit only if they don't shoot up too high, so in the practical arts the leaves and

117 Clausewitz, 'On the Life and Character of Scharnhorst', p. 99.

118 See Clausewitz, 'The Germans and the French', in Clausewitz, *Writings*, p. 57. See also, Roger Parkinson, *Clausewitz: A Biography* (New York: Stein and Day, 1979), p. 94.

119 Carl von Clausewitz, 'Notes on History and Politics', in Clausewitz, *Writings*, p. 270. A similar concern is highlighted when Clausewitz asks: what is the value of 'obscure, partially false, confused and arbitrary notions? Very little – so little that they have made theory, from its beginnings, the very opposite of practice, and not infrequently the laughing-stock of men whose military competence is beyond dispute'. Clausewitz, *On War*, II, 5, p. 198.

120 *On War*, II, 5, p. 197.

121 Clausewitz, 'On the Life and Character of Scharnhorst', p. 104.

> flowers of theory must be pruned and the plant kept close to its proper soil – experience.[122]

Clausewitz frequently employed a deductive approach to explore the boundaries of theoretical concepts: historical accounts would never entirely suffice because, as Scharnhorst had recognised, 'experience without philosophy is devoid of meaning'.[123] Book 8, Chapter 2 provides an excellent example. Logically, he states, it would appear that the overthrow of the enemy should govern all war plans. That all wars do not reach such extremes might be explained by limiting factors inherent in the 'war-machine', but these are deemed insufficient to 'span the gap between the pure concept of war and the concrete form'. Reconciliation is only achieved when it is accepted that 'war and its forms result from ideas, emotions, and conditions prevailing at the time' in relation to which 'strictly logical reasoning often plays no part at all and is apt to be a most unsuitable and awkward intellectual tool'. The logical absolute retains its usefulness as a 'general point of reference', but reality has its own necessity, determined by particular conditions that always makes war 'a matter of degree'. All the wars fought from Alexander to Bonaparte cannot be rejected because they do not equate to the theoretical ideal; rather, the divergence must be explained.[124] In this way, Clausewitz's use of abstract reasoning was not a consequence of intellectual pretension, but a means of revealing fundamental truths.

So, as the counterweight to pure theory, 'experience' for Clausewitz was a broad notion encompassing two central aspects, both concerned with the reality of war: history and personal experience. He held that history provided the theorist with a vast reservoir of experience that 'counts for more than any amount of abstract truths'.[125] As Paret and Moran note, Clausewitz believed that, 'Without the instrument of history, theory should not be constructed … History not only tested and validated his theories, it gave rise to some of them'.[126] However, for Clausewitz, this meant much more than familiarity with a few history books. It was not simply the case, as Livy had it, that history is a 'record of the infinite variety of human experience plainly set out for all to see'.[127] As Howard notes, the 'exercise of history had itself to be an exercise in critical judgement'.[128] A detailed basis of such critical historical study is masterfully outlined in Book 2 of *On War*. For Clausewitz, deep study of the past provides the theorist with something

122 Clausewitz, 'Author's Preface', in *On War*, p. 69.

123 Strachan, 'Dialectics', p. 38.

124 *On War*, VIII, 2, pp. 700–703.

125 *On War*, II, 5, p. 191.

126 Peter Paret and Daniel Moran, 'Introduction to Part One', in Clausewitz, *Writings*, p. 3.

127 Livy, *The Early History of Rome*, trans. Aubrey de Sélincourt (London, Penguin, 2002), p. 30.

128 Howard, *Clausewitz*, p. 31.

approaching recorded reality and thus history becomes the starting point for theoretical analysis. It also serves to confirm genuine advances in understanding.

The other potential source of observations on war's reality was of course first-hand experience. It is certainly rare to find thinkers on war who combine the qualities of both soldier and philosopher. Clausewitz's experiences added to the depths of his comprehension. Many sections of *On War* and his historical studies draw on, implicitly, his own recollections of campaigns.[129] He also suggested that only those who have experienced war can truly be in a position to emphasise aspects that may evade the armchair theorist. For instance, the danger, fear and friction that affect war so greatly may not be directly apparent to the mere spectator or from the comfort of the ivory tower. As Clausewitz pithily remarks, 'in one's library these elements are not known'.[130] That these factors play such an integral role in his theory owes a great deal to his first-hand knowledge. As he notes, 'theorists who have never learned to generalise from experience, are impractical and even ridiculous: they teach only what is already common knowledge'.[131] The difficult task, however, was translating these experiences into theory and expressing them through coherent concepts.

So, Clausewitz conceived of a near ceaseless interaction between reasoning and experience. They should be in constant dialogue with one another as neither is much use in isolation. The practical soldier in him led Clausewitz to favour practice, but if theory was to improve men's understanding of war then one had to make sense of the reality through philosophy: 'Analysis and observation, theory and experience must never disdain each other; on the contrary, they support each other'.[132] The two are mutually supportive as practice requires understanding, whilst theory should not render itself useless by becoming detached from reality (a major modern problem in relation to nuclear strategy).[133] The importance of resolving this tension is revealed in a heated outburst in *On War* when Clausewitz states that theory,

> only needs intelligent treatment to make it conform to action, and to end the absurd difference between theory and practice that unreasonable theories have

129 His study of the 1812 Campaign is more overtly personal in character. Clausewitz, 'Campaign of 1812', pp. 113–204.

130 Clausewitz, 'Campaign of 1812', p. 166.

131 Clausewitz, 'Campaign of 1812', p. 139.

132 Clausewitz, 'Author's Preface', in Clausewitz, *On War*, p. 69.

133 During the Cold War, civilian 'systems analysts' with no direct experience of military affairs nor historical empirical evidence on which to base their analyses, formulated strategies for the threat or use of nuclear weapons. Their theories, perhaps understandably, represent extremes of the speculative approach Clausewitz criticised. They achieved their conclusions only by radically simplifying reality, concentrating on statistically tractable problems, positing a-cultural rational actors, downplaying emotional and psychological factors and making assumptions that might not pertain in actual scenarios. Hedley Bull, 'Strategic Studies and its Critics', *World Politics*, Vol. 20, No. 4, July 1968, pp. 600–601.

> so often evoked. That difference, which defies common sense, has often been used as a pretext by limited and ignorant minds to justify their congenital incompetence.[134]

There is, however, still one area of reality that theory must encompass if it is to retain its relevance. History not only revealed to Clausewitz the incredible variety of wars in the past, but opened his eyes to the vast potential of the future. This sentiment is revealed in a section in which he is speculating on whether the sort of limited wars, common throughout history, will recur: 'we should be bound to say that in spite of our theory [of absolute war] there may even be other wars of this kind in the next ten years, and that our theory, though strictly logical, would not apply to reality'.[135] Clausewitz was not only concerned with wars in recorded history or that he had personally experienced, but also with those yet to occur.

Material versus moral factors

The emphasis Clausewitz placed on the importance of experience led him to another crucial issue. He notes how, previously, the study of war concerned 'only the total body of knowledge and skill that was concerned with material factors'.[136] As such, war 'would hardly provide a scientific problem for a schoolboy'.[137] Most earlier theorists had directed their attention to physical matters such as issues of troop numbers, supply, geometry and even geology.[138] These theories, he explains, were about as useful 'to combat as the craft of the swordsmith to the art of fencing. It did not yet include the use of force under constant conditions of danger … nor the efforts of spirit and courage to achieve a desired end'.[139] Some of these thinkers, such as Saxe and Lloyd, had placed significant emphasis on the importance of psychological forces, but only in a haphazard manner.[140] For instance, in his *Reveries,* Saxe noted how, in war, 'the solution lies in human hearts' and that this matter is the 'most important, the most learned and the most profound, of the profession of war'.[141] Nevertheless, his general insights on that

---

134 *On War*, II, 2, p. 164.

135 *On War*, VIII, 2, p. 702.

136 *On War*, II, 2, p. 153.

137 *On War*, III, 1, p. 208.

138 Clausewitz discusses this in relation to Mathieu Dumas who 'discovered that the fundamental principle lay in possession of the higher ground. By way of numerous half truths and doubtful conclusions this led to a picturesque system of geological analogies. The neighbours of the clouds, the highest mountain peaks, rule the land below them, while the rivers become the couriers of their might. The art of war seems to arise geologically from the bowels of the earth'. Clausewitz, 'On the Life and Character of Scharnhorst', p. 103.

139 *On War*, II, 2, p. 153.

140 Paret, *Clausewitz*, pp. 157–8.

141 Saxe, *Reveries*, p. 18.

subject are not given a firm theoretical basis, no matter what excuses he gave for the 'irregularity of the arrangement, as well as the inelegance of the style' of his work.[142]

Progressing beyond these limited systems, Clausewitz emphasised that an observation of reality revealed that 'moral forces' are fundamental to war and must be incorporated in theory. Previous theories had directed their inquiries predominantly 'toward physical quantities, whereas all military action is intertwined with psychological forces and effects'. He argued that, while material factors may yield to the application of reason, war, as a human social activity, is universally bound up with intangible factors where the 'rules dissolve into nothing but vague ideas'.[143] Just as Goethe revealed a man's physical weakness to be relative by describing how he can be 'so enraged by an insult that he takes on six opponents and overcomes them,'[144] so Clausewitz stressed that analysis of physical factors alone was insufficient. He observed that, 'All these and similar effects in the sphere of mind and spirit have been proved by experience: they recur constantly, and are therefore entitled to receive their due as objective factors. What indeed would become of a theory that ignored them?'[145] It was therefore the task of theory to 'analyse emotional forces of all kinds: the psychology of the individual and the psychology of the group'.[146]

Clausewitz thus developed concepts that he felt best expressed these factors, such as genius, passion, morale, courage and so forth. Their incorporation in theory mirrors what Clausewitz felt was the existing dichotomy between theory and practice: theory had a tendency to focus on material factors as distinct from practice, which was inevitably pervaded by moral forces. Yet, he acknowledged this was a difficult task: conceptualisation of these factors would always be limited because they relate to values that 'can only be perceived by the inner eye, which differs in each person'.[147] He did not believe that such aspects of war could be quantified, but that they must be studied as far as inherent limitations allow.

Moreover, any theoretical observation on such issues must be based on *actual experience* rather than fanciful speculation: 'no theorist, and no commander, should bother himself with psychological and philosophical sophistries'.[148] Perhaps Clausewitz had in mind here earlier thinkers, such as Berenhorst, who had overplayed moral factors or expressed them in a manner inimical to sound analysis. The recognition of such factors should not, he believed, preclude the articulation of clear concepts. He did not entirely reject the possibility of a scientific approach,

---

142 Saxe claimed that he wrote his work while sick with fever.

143 *On War*, II, 2, pp. 156–7.

144 Johann Wolfgang von Goethe, *The Sorrows of Young Werther* (London: Penguin, 1989), p. 62.

145 *On War*, II, 2, p. 158.

146 Paret, *Clausewitz*, p. 373.

147 *On War*, II, 2, p. 158.

148 *On War*, II, 2, p. 158.

even when such intangibles were involved, although he admitted that 'theory becomes infinitely more difficult as soon as it touches the realm of moral values'.[149]

Just as theory had to reconcile ideas and practice, so too did it need to reconcile material and moral forces. Clausewitz held that the two 'cannot be separated' and should be conceived as 'an organic whole which, unlike a metal alloy, is inseparable by chemical processes'. Moreover, he believed moral forces could be the most decisive in war: 'One might say that the physical seem little more than the wooden hilt, while the moral factors are the precious metal, the real weapons, the finely-honed blade'.[150] Napoleon had reached a similar conclusion when he had stated that 'the moral is to the physical as three to one'.[151] One of the principal dangers of abstract theorising was its tendency to ignore such intangible yet often decisive elements because they do not translate neatly into doctrinal precepts. The emphasis on psychological factors only reinforced Clausewitz's belief that theory should not be dogmatic – for instance, what theory could actually teach courage in the face of danger? All theory could do is acknowledge their importance.

The universal versus the particular

Aron has observed that 'strategic thought draws its inspiration each century, or rather at each moment of history, from the problems which events themselves pose'.[152] And so, Enlightenment theorists, in their attempts to create a timeless science of war, focused too heavily on their own times and paradoxically sacrificed universal application for contemporary relevance. The great Napoleonic campaigns dominated the minds of military thinkers. Those cataclysmic events not only represented a whole new form of war, but for active officers like Clausewitz, analysis of them was powerfully linked to action. Theory had to be closely linked to reality not only in the interest of methodological rigour, but also for the sake of national survival. As Howard notes, Clausewitz 'had the practical man's horror of abstractions that could not be directly related to the facts of the situation'.[153]

This was undoubtedly so, but Clausewitz was equally concerned to establish those aspects of war that were immutable. As Paret states, for Clausewitz, 'to devise effective strategy and tactical measures mattered far less to him than to identify the *permanent elements of war* and come to understand how they function.[154] He intended his work to last, that 'it would not be forgotten after two or three years, and that possibly might be picked up more than once by those who

149 *On War*, II, 2, p. 157.

150 *On War*, III, 3, pp. 216–17.

151 Quoted in Rupert Smith, *The Utility of Force: The Art of War in the Modern World* (London: Penguin, 2005), p. 242.

152 Raymond Aron, 'The Evolution of Modern Strategic Thought', in Alistair Buchan (ed.), *Problems of Modern Strategy* (London: Chato and Windus, 1970), p. 25.

153 Howard, *Clausewitz*, pp. 2–3.

154 Paret, 'Clausewitz', in Paret, *Makers*, p. 187. Emphasis added.

are interested in the subject'.[155] As Hugh Smith notes, Clausewitz believed there existed 'a universal element in war: common and enduring features that escape change and must not be lost from view'.[156] The problem of reconciling these two poles, the universal and the particular – objective and subjective knowledge – was not a novel problem and has occupied the minds of many philosophers throughout history. But, as Clausewitz understood, recognition of the problem was not a sufficient condition for its resolution.

Clausewitz had understood that too great an emphasis on the universal leads to the Scylla of banal generalities and dull truths, whereas emphasis on the particular led to the Charybdis of mindless trivialities and sterile pedantry. As we have seen, his fear of descending into dogmatism restrained him from making theoretical claims of relevance only to the particular circumstances under study and which would cause one to 'drown in trivialities'.[157] He was also acutely aware of the limits of theory and that any attempt to establish timeless truths was fraught with danger. As he recognised, most military theorists had been over-impressed by the Napoleonic period, drawing hasty conclusions from the principles it appeared to emphasise: as he warned, 'Whoever relies purely on the perspectives of his own times is inclined to treat what is most recent as best'.[158]

It could be argued that Clausewitz very nearly fell into the trap of particularism. Two key factors perhaps explain how he avoided this pitfall. First, it would be wrong to view the impact of the events of his times in a unitary fashion. Rather, the period he lived through is perhaps best conceived of as being split into two principal stages, both with unique features. As Creveld explains, 'Born and raised in one set of axioms, rules, theories, beliefs, he saw it abruptly shattered by defeat and replaced by another superior to it. It was this fact … which enabled him to compare both sets [and] contrast them'.[159] This variety of experience forced him to try to reconcile such divergent periods and consider their differences and similarities, their contradictions and commonalties. Of course, simply witnessing this transition was not sufficient to give rise to Clausewitz's mature theories. His contemporaries drew quite different conclusions: in particular, that the decisive Napoleonic campaign represented the apotheosis of war. A wider perspective enabled Clausewitz to avoid this trap.

Second, the developing historicist school underlined for him the particularity of different epochs and cultures. As noted, he had great sympathy with Herder's view that different ages should be assessed in their own right, indeed, so much so that Clausewitz claimed that every age 'would have held to its own theory of war'.[160] Of course, an extreme version of historicism can lead to a form of

155 Clausewitz, 'Author's Comment', in *On War*, p. 72.
156 Smith, *On Clausewitz*, p. 184.
157 Clausewitz quoted in Smith, *On Clausewitz*, p. 173.
158 Quoted in Strachan, 'Dialectics', p. 39.
159 Creveld, 'Eternal Clausewitz', pp. 46–7; Parkinson, *Clausewitz*, p. 310.
160 *On War*, VIII, 3B, p. 717.

dusty antiquarianism or relativist belief that the singularities of unique periods serve to make attempts at abstract conceptual generalisations futile. Clausewitz never went this far. He was not a relativist and did not dismiss the possibility of comparison across ages. He believed that the analyst must liberate himself from the fashions and constraints of his own age to discover those elements that are comparable and present in every war.[161] The historicist perspective acted as a check on the tendency to universalise his present – the Napoleonic experience – and compelled him to view it in *its own unique context* and 'helped him avoid the error of defining Napoleonic war as the 'correct' war'.[162] An awareness of the incredible variety of wars throughout history forced him to engage in a difficult process of distinguishing what is *essential* from what is *incidental*.

This is not to suggest that he always managed to bridge this divide. Sections of *On War* 'where he comes closest to offering advice of the how-to-do-it-variety … are most often regarded as hopelessly obsolete'.[163] Nevertheless, he was determined to develop a theory that was applicable in his own time as well as others and that did not offer bland platitudes or overtly disagree with experience. It is apt to quote a passage by Clausewitz that encapsulates his thinking on this issue:

> Philosophy teaches us to recognise the relations that essential elements bear to one another, and it would indeed be rash from this to deduce universal laws governing every single case, regardless of all haphazard influences. Those people, however, who 'never rise above anecdote,' … never get down to the general features that govern the matter. Consequently their findings will never be valid for more than a single case; indeed they will consider a philosophy that encompasses the general run of cases as a mere dream.[164]

*

In the course of writing *On War*, Clausewitz attempted to reconcile all these apparently contradictory elements of theory. We have seen how he explored the extremes of these methodological dualities and sought their resolution by appropriating the most valuable aspects of each and fusing them to create a theoretical dualism, like a negotiated agreement between two conflicting parties. It is in the trinity that the fruits of these labours are to be found. Clausewitz seeks to provide clear concepts derived from both reason and experience, which fuse material and ideational forces and that are present in every war, yet manifest themselves according to the circumstances. His central conclusions are intended not as doctrinal prescriptions, but as guides to understanding. The trinity represents the complete integration of these dualisms.

161 Paret, *Clausewitz*, p. 153.
162 Paret, *Understanding*, p. 152.
163 Creveld, 'The Eternal Clausewitz', p. 35.
164 *On War*, VI, 6, p. 448.

Even if we do not subscribe fully to the content of Clausewitz's theory, what should not be doubted is that he provided a standard of excellence in terms of developing a methodology which could enable the formulation of universal theory. It is a methodology that confronts the theorist with the most intractable problems in social science and deals maturely with many of the issues that continue to vex the minds of theorists today: the nature of the relationship between theory and practice, between the material and psychological, between the universal and particular. It might even be argued that the lucid way Clausewitz tackled these problems is superior to the often convoluted language of many modern political scientists.

Few military thinkers have approached the subject with such dedication, wide-ranging experience or intellectual integrity, combined with a strong desire to avoid 'an ostentatious exhibition of ideas'[165] as so often accompanies intellectual vanity. Clausewitz faced his theoretical problems almost with what at times appears to be a slight sense of desperation, as if his life depended on the successful resolution of some nagging logical contradiction between theory and reality. In some senses this was true. Late in his life, Clausewitz grew increasingly aware that he would probably not achieve lasting renown on the battlefield. And so he devoted his final years to an area in which he would ultimately achieve lasting influence: theory. Even those who disagree with the substantive conclusions of his theory might at least recognise that Clausewitz set the standard in tackling a subject that has evaded the grasp of many a great mind.

165 *On War*, II, 5, p. 197.

Chapter 3
Context and Circumstance

A cursory glance through any number of military history books reveals that most authors feel compelled to say something, whether succinctly or at length, about the context within which a particular battle or war took place. The impulse to describe the broad setting in which events occurred is, it seems, an almost natural urge. Return to some of the earliest classical historians and this is apparent. Thucydides, for instance, begins his *History of the Peloponnesian War* with a penetrating account of the political situation that culminated in the conflict between Sparta and Athens.[1] Without such information, meaning and explanation pay a heavy price. As a result, newcomers feel compelled to consult other accounts to gain a bearing and better understand the events described. A parallel can be made here with the military commander who attempts to use force with little understanding of their wider environment, concerned only with the immediate problem of defeating the enemy. Moreover, the issues are more closely linked than may appear at first sight. Sound history teaches commanders that outcomes in war are shaped by the wider conditions in which they take place. It is the task of theory to conceptualise this idea.

In his magisterial history of *The Franco-Prussian War*, Michael Howard makes it clear that to understand apparently isolated military events, we must seek to comprehend how those events are connected to wider forces.[2] An account of the former may suffice to *describe* the course of war, but certainly not *explain* it. Simply observing a particular event and the elements that constitute it is only a first step, albeit an important one, to understanding. The implication is that the behaviours and characteristics of armies, weapons technology and supply systems are as they are *for a reason*. They are all products of an age with its own unique features at a certain point in time and place. If we fail to inquire as to why things are as they are, then our understanding of events will be necessarily parochial. The broader perspective will usually hold 'the key for a full understanding of the fighting'.[3] This is a perspective embodied in the 'war and society'[4] or 'New Military History' school which, as Geoffrey Wawro explains, 'does not skimp on

1 Thucydides, *History of the Peloponnesian War*, trans. Rex Warner (London: Penguin, 1972).

2 Michael Howard, *The Franco-Prussian War* (Abingdon: Routledge, 2007), pp. ix–x.

3 Peter Paret, *Understanding War: Essays on Clausewitz and the History of Military Power* (Princeton: Princeton University Press, New Jersey, 1993), p. 1.

4 Jeremy Black, *Introduction to Global Military History: 1775 to the Present Day* (Abingdon: Routledge, 2005), pp. 270–71.

technology, generalship or battles, but takes care to place war in its larger social, political, economic and cultural context'.[5]

These points may come across as basic common sense, but it is surprising how often this crucial aspect of war is neglected by historians, strategists and theorists. Clausewitz pointed to this evasive inclination in a letter to a colleague with regard to the centrality of comprehending the political conditions of any war: 'That it is essential to see the matter in this way, that the point of view is almost self-evident if we only keep the history of war in mind, scarcely needs proof. Nevertheless, it has not been fully accepted'.[6] Gray has also noted that, 'since, following Clausewitz, war cannot be approached as an autonomous activity, if it is considered bereft of context it becomes literally senseless'.[7]

So, how are we to conceive of the concept of context, how and why does it contribute to our understanding of war and what is its role in the trinity? Context is a vital aspect in the framework as it provides the 'setting' in which the three primary tendencies – passion, chance and politics – are manifested in reality through secondary level subjects. It can be conceived as a tertiary level in the framework, underlying, intertwined with and inseparable from the others. The chapter begins by exploring the various ways in which context is understood by Clausewitz. The following section considers the concept in detail and offers a clearer articulation of context as a central component of a trinitarian theory of war. The chapter concludes by outlining some of the prominent substantive dimensions of context by way of historical example.

Clausewitz on context

Clausewitz engages with this concept in a number of ways and in relation to a variety of issues. This provides us with differing routes into the subject and allows us to reveal how Clausewitz employed the idea. That context was of great importance for Clausewitz is unmistakable if one explores, in detail, *On War* and his other writings. Indeed, as Gat asserts, 'throughout his life … this view of war within the context of its particular social and political reality was fundamental to Clausewitz's historical and theoretical outlook'.[8] Aron also notes that Clausewitz 'could not fail to discover or observe the diversity of wars

5 Geoffrey Wawro, *Warfare and Society in Europe, 1792–1914* (London: Routledge, 2000), p. ix.

6 Carl von Clausewitz quoted in Paret, *Understanding*, p. 127.

7 Colin S. Gray, *Another Bloody Century: Future Warfare* (London: Weidenfeld & Nicolson, 2005), p. 56.

8 Azar Gat, *A History of Military Thought: From the Enlightenment to the Cold War* (Oxford: Oxford University Press, 2001), p. 218.

according to their times'[9] given his own personal experiences, combined with his historical study of war.

In a general sense, the importance of conditions and circumstances is frequently expressed in relation to more practical military issues. When Clausewitz considers strategic and operational principles, the analysis is almost always supported by accompanying caveats. These remarks – described by Beyerchen as a 'maddening maze of qualification'[10] – amount to a recognition of the importance of unique conditions which allow for either the moderation or even abandonment of action according to certain principles.[11] As Clausewitz states, 'Many roads lead to success and they do not all involve the opponent's outright defeat ... the choice depends on circumstances'.[12] For instance, Clausewitz explains how a certain style of fighting derived from a successful experience fighting one war 'can easily outlive the situation that gave rise to it; for conditions change imperceptibly'. This was a fact devastatingly revealed at Jena in 1806 when Frederick the Great's 'oblique order' proved wholly unsuited to battle against Napoleon: an example of 'the most extreme poverty of the imagination to which routine has ever led'.[13] While such operational follies are important and litter the history of war (popularly expressed in the aphorisms that 'generals are always fighting the last war' or that 'nothing fails like success'), this section will concentrate on more fundamental grounds for Clausewitz's belief in the centrality of context.

History and historicism

The clearest expression of the importance of contextual understanding is to be found in Clausewitz's work which focuses on historical matters. We have already noted the importance of history to Clausewitz's theoretical approach, but here we need to understand its crucial influence in developing his sensitivity towards contextual conditions. Clausewitz's reading of Herder and Möser, and other writers of the Historicist school, taught him to appreciate the individuality of different historical epochs and cultures. It was an approach to history which had at its heart the idea that past societies should be interpreted on their own terms, not simply as earlier inferior stages on the long march toward higher civilisation, as the dominant progressive Enlightenment perspective had it. Historicism placed emphasis on understanding the conditions that shaped the culture, politics and life of earlier societies.

9 Raymond Aron, *Clausewitz: Philosopher of War*, trans. C. Booker and N. Stone (New Jersey: Prentice-Hall, 1985), p. 41.

10 Alan Beyerchen, 'Clausewitz, Nonlinearity and the Unpredictability of War', *International Security*, Vol. 17, No. 3, Winter, 1992, p. 84.

11 Carl von Clausewitz, *On War*, trans. and ed. Michael Howard and Peter Paret (New York: Everyman's Library, 1993), I, 2, p. 111.

12 *On War*, I, 2, p. 107.

13 *On War*, II, 4, p. 180.

Perhaps the best example of Clausewitz's historicism is to be found in his historical survey in Chapter 3B, Book 8. Clausewitz explains that different historical societies 'all conducted war in their own particular way, using different methods and pursuing different aims'.[14] After impressionistically reviewing different periods, he concludes by stating that, 'We wanted to show how every age had its own kind of war, its own limiting conditions, and its own peculiar preconceptions … It follows that the events of every age *must be judged in the light of its own peculiarities*'.[15] These ideas had been apparent to Clausewitz in his early twenties when writing about the Thirty Years War. Criticising scholars who saw those wars as nothing but formless and brutish struggles, Clausewitz emphasised that the men of the age 'acted in accordance with their economic and technological condition, their political and religious concerns, and their psychology'.[16]

Paret, in an essay considering Clausewitz's historical method, draws attention to these ideas. He describes how Clausewitz, in his efforts to understand the great changes wrought by the French Revolution, felt compelled 'to fit the Revolution into the larger processes of European history' because it could be 'accurately interpreted only if the conditions preceding it were taken in to account'.[17] Similarly, Clausewitz recognised that the military institutions of the *ancien régime* were rendered almost obsolete because their 'forms and means were no longer appropriate to the changed times and political conditions'.[18] This sentiment is also expressed in his 'Observations of Prussia in Her Great Catastrophe', relating to the humiliating defeat suffered by Prussia in 1806, in which he states that 'the machinery of government was desiccated, decrepit, and *entirely unsuited to the times*'.[19] These examples reveal the extent to which Clausewitz always strove to explain events – both historical and contemporary – in relation to prevailing conditions, as opposed to judging them according to some universal standard; a fault Clausewitz recognised in alternative approaches, in particular those with a teleological bent.

Dangers of the teleological approach

Paret notes how Clausewitz criticised other writers for their teleological view of history, stating that, 'History in the service of a philosophic worldview as

14 *On War*, VIII, 3B, p. 709.

15 *On War*, VIII, 3B, p. 717. Emphasis added.

16 Paret, *Understanding*, p. 211.

17 Paret, 'Introduction', in Carl von Clausewitz, *Historical and Political Writings*, ed. and trans. Peter Paret and Daniel Moran (New Jersey: Princeton University Press, 1992), p. 8.

18 Clausewitz, 'On the Life and Character of Scharnhorst', in Clausewitz, *Writings*, p. 102.

19 Clausewitz, 'Observations on Prussia in Her Great Catastrophe', in Clausewitz, *Writings*, p. 41. Emphasis added.

Hegel encapsulated it, for instance, would not serve Clausewitz's purpose'.[20] Hegel's philosophy is a good instance of the radically contrasting approaches to history that exist and underlines the extent to which Clausewitz's historicism and objectivity are by no means default positions. Hegel, in his *Lectures on The Philosophy of History*, seeks to reveal that history is essentially the manifestation of a rational process; the gradual unveiling of Reason or Spirit as Freedom, which had supposedly reached its apotheosis in the Prussian state of his times.[21] History is presented as an impersonal process that rejects what is not pertinent to its underlying *telos*, leaving only that which correlates to reason as 'Actual'. Even dominant Western perspectives today contain a large measure of an almost unconscious historical teleology, based as they often are on broad Enlightenment ideals of progress and perfectibility.[22]

A very similar view – almost the military theoretical equivalent of Hegel – is apparent in the work of Clausewitz's contemporary, Jomini. He considered all earlier periods merely preparatory to the present, which had revealed permanently valid principles – 'fixed laws resembling those of the positive sciences'[23] – as embodied in the campaigns of Napoleon (mirroring Hegel's eulogy to the Prussian state) and can be summarised as: 'to throw by strategic movements the mass of an army, successively, on the decisive points of a theatre of war'.[24] Clausewitz rejected this view because he believed that such principles were 'dependent on the circumstances of the time ... which might not recur under altered conditions'.[25] Jomini held that a 'reality lies beneath the superficial chaos of the historical moment in enduring and invariable principles, like those of gravitation and probability'.[26] His principles of the 'decisive point' and 'interior lines' were the correlative of Hegelian 'Actuality'.

In a manner typical of most teleological approaches, Jomini judged past wars, such as Frederick's campaigns, on the unhistorical basis of principles that had been revealed in their correct form in the present. Even if he recognised differences in the character of warfare and that 'the specific application of principles would vary,'[27] this did not prevent him from judging different periods – 'from Scipio and Caesar to Napoleon' – according to one standard.[28] Past defeats could be explained by a failure to adhere to the principles he had revealed: '*Le système de L'Empereur*

20 Paret, *Understanding*, p. 132.

21 Georg Hegel, *Lectures on the Philosophy of History* (Whitefish: Kessinger, 2004).

22 See John Gray, *Black Mass: Apocalyptic Religion and the Death of Utopia* (London: Penguin, 2007).

23 Antoine Henri de Jomini, *The Art of War*, trans. G. H. Mendell and W. P. Craighill (Texas: El Paso Norte Press, 2005), p. 258.

24 Jomini, *Art of War*, p. 55.

25 Paret, *Understanding*, p. 137.

26 John Shy, 'Jomini', in Peter Paret (ed.), *Makers of Modern Strategy: From Machiavelli to the Nuclear Age* (Oxford: Oxford University Press, 1986), p. 148.

27 Shy, 'Jomini', p. 154.

28 Quoted in Gat, *Military Thought*, p. 124.

Napoléon présente une application constante de ces principes invariable'.[29] Clausewitz criticises such approaches when he states that, 'Whoever relies purely on the perspective of his own times is inclined to treat what is most recent as best – and he finds it impossible to deal with what is exceptional or out of the ordinary'.[30]

This brief examination of Jomini's thought reveals that the study of history does not automatically entail an appreciation of the importance of conditions. In Jomini's case, the past essentially represented a handy store of examples that could be used to prove his principles. Under Jomini's hand, theory would shape history, rather than history shaping theory. Clausewitz, in opposition to Jomini, held to the latter approach and developed a theory that was more sympathetic to historical conditions and that, somewhat paradoxically, would be more universally relevant than Jomini's supposedly 'scientific' system. As Gat has made clear, Clausewitz rejected Jomini's approach because it 'ignored the living reality of war … and the unique conditions of each particular case'.[31]

That the Jominian approach is not without its more modern equivalents is revealed in Basil Liddell Hart's *Strategy*.[32] In a fashion similar to that of Jomini, Liddell Hart – inspired by a moral indignation toward the form of 'total war' witnessed during World War I and relying on the eighteenth-century precedent – prescribed operations on 'the line of least expectation'.[33] As Gat notes, his search for the universal principle of the indirect approach as the basis of all decisive successes 'was marred by his unhistorical and naïve disregard for the particular conditions … which had shaped each period's way of doing things'.[34] Strachan notes that of course, 'the line of least expectation may in fact be that of supposed greatest expectation'.[35] Again, deductive theory was allowed to usurp inductive, critical historical analysis, thus perverting historical truth for its own ends.

Context in theory and historical method

Despite, in the previous chapter, having discussed Clausewitz's approach to theory and history it is worth exploring these ideas further, because contained in *On War* are valuable clues as to the importance he attached to context. For instance, in his chapter on 'Method and Routine', Clausewitz rejects the notion of scientific 'laws' being applicable to war 'since no prescriptive formulation universal enough to deserve the name of law can be applied to the constant change and diversity of the

29 Quoted in Shy, 'Jomini', p. 152.

30 Quoted in Hew Strachan, 'Clausewitz and the Dialectics of War', in Hew Strachan and Andreas Herberg-Rothe, *Clausewitz in the Twenty-First Century* (Oxford: Oxford University Press, 2007), p. 39.

31 Gat, *Military Thought*, p. 125.

32 Basil Liddell Hart, *Strategy* (New York: Meridian, 1991).

33 Liddell Hart, *Strategy*, p. 327.

34 Gat, *Military Thought*, p. 682.

35 Strachan, *European Armies*, p. 3.

phenomena'.[36] Indeed, it is this idea that explains many of the arguments Clausewitz puts forward with regard to the limits and purpose of theory. For example, he explains that earlier efforts to produce a positive theory of war resulted from a 'failure to take adequate account of the endless complexities involved'. Theory cannot aim to become a commander's manual for action intended 'to serve as a guide which at the moment of action lays down precisely the path he must take' because each situation presents different circumstances; due to changing contexts, all theory can hope to provide is a form of intellectual foundation, 'not a model for the art of war'. If conditions throw up different strategic imperatives then theory, to remain relevant, must be capable of accommodating this.[37]

In his chapter on 'Historical Examples', we gain a further indication of the centrality of context to his theoretical approach. The relevant section concerns the use of historical examples as proof for certain propositions.[38] Clausewitz notes how some people believe that in using examples they can substitute quantity for detail: 'instead of presenting a fully detailed case, critics are content merely to *touch on* three or four, which give the *semblance* of strong proof'. This, in his view, is mistaken: 'an event that is lightly touched on … is like an object seen at great distance … it looks the same from every angle'[39] and thus cannot offer conclusive proof. For this reason, a single thoroughly detailed event is to be preferred over a number of limited cases. This was a conclusion he had reached as early as 1812 when, in his written advice to the Crown Prince he advised that, 'The detailed study of a few individual engagements is more useful than the general knowledge of a great many campaigns'.[40]

The reasoning behind these assertions is revealed when he explains why examples drawn from recent history will be more reliable and susceptible to considered analysis. The problem is essentially one of context and relates to an issue raised earlier. Clausewitz states that, 'not only were conditions different in more distant times … but military history is bound with the passage of time to lose a mass of minor elements and details … It loses some element of life and colour, like a picture that gradually fades and darkens'. He goes on to explain that the study of more recent history is to be preferred because 'the conditions of modern warfare … bear a considerable resemblance to those of the present day … The further back one goes, the less useful history becomes, growing poorer and barer at the same time'.[41]

Clausewitz's comments are not to be understood as placing the lessons of modern history over and above ancient history in an objective sense: all periods

36 *On War*, II, 4, p. 176.

37 *On War*, II, 2, pp. 154–61.

38 *On War*, II, 6, pp. 200–201.

39 *On War*, II, 6, p. 202.

40 Carl von Clausewitz, *The Principles of War*, trans. and ed. Hans W. Gatzke (New York: Dover Publications, 2003), p. 68.

41 *On War*, II, 6, p. 203.

and wars are *intrinsically* equally valuable to the formulation of a general theory of war, just not to the inevitably timebound individual theorist. The determining conditions of the age under study will – due to the lack of available evidence and the theorist's poor understanding of very different times – be improperly grasped and, therefore, establishing proofs or making judgements on the correct use of means will be next to impossible. The critical historian of ancient warfare is largely presumed to be 'in no position to evaluate the relevant events correctly, nor to apply them to the wholly different means we use today'.[42] A similar concern explains Clausewitz's criticism of Machiavelli, which he details in an 1809 letter to the philosopher Fichte: 'the art of war of the ancients attracted him too much, not only its spirit, but also in all of its forms'.[43]

It is this line of argument that helps explain, and ultimately reject, a criticism that has been directed at Clausewitz: that he is over-reliant on historical examples drawn only from the relatively recent past and, in particular, the Napoleonic period. Such a claim not only downplays the extent of his broad historical knowledge, but also fails to grasp Clausewitz's *conscious decision* to derive most of his propositions from recent experience. Due to his conviction that meaningful conclusions can only be established based on a deep understanding of prevailing conditions, he was not willing to succumb to 'vanity and quackery' in order to provide a cover of respectability – 'sheer decoration' – to propositions that would almost inevitably lack validity.[44]

Context and the trinity

And so we return to the trinity. How do the many ideas described above find their final expression in his theoretical synthesis? It would be perhaps odd if, after emphasising such issues in his wider work, Clausewitz did not incorporate them, in some shape or form, in the trinity.

The answer is that context *is* expressed in the trinity, but in a manner that does not sufficiently convey the theoretical importance of the concept. In fact, when the meaning is grasped, it becomes clear that the issue of context is presented in the very first line of the section outlining the trinity. Clausewitz states that 'war is more than a true chameleon that slightly adapts its characteristics to the given

42 *On War*, II, 6, p. 204.

43 Carl von Clausewitz, 'Letter to Fichte', in Clausewitz, *Writings*, p. 281. It would be unfair however, to state that Machiavelli did not at least recognise the potential pitfalls of over-reliance on ancient practice. In the Preface to Book II of *The Discourses*, Machiavelli explains that it is perhaps dangerous because, 'The whole truth about olden times is not grasped' and he realises that he may be amongst those 'who deceive themselves if in these my discourses I have praised too much the days of the ancient Romans'. Machiavelli, Niccolò, *The Discourses*, trans. Leslie J. Walker (London: Penguin, 2003), pp. 265 and 268.

44 *On War*, II, 6, p. 204.

case'.[45] What could be mistaken for a throw-away comment or handy metaphor, in fact embodies much of what has been explained above. Clausewitz was a master of metaphor and few were included that failed to convey accurate meaning.[46] Contrary to popular belief, chameleons do not change colour according to their surroundings, but rather as a result of their inner physiological condition. Yet, it is clear that Clausewitz has in mind the popular conception of a chameleon's ability to blend with its surroundings. So, what does this apt, if scientifically inaccurate, analogy tell us?

In comparing war to a chameleon Clausewitz is suggesting that war's outer appearance – the 'face of war' – is in constant flux, even though its inner nature – the trinity – remains unchanged. War is 'more than' a chameleon because he explains how the three tendencies will constantly change in their relationship to one another.[47] Here, the scientifically correct chameleon would perhaps be a more appropriate analogy. The three tendencies represent permanent, shifting features of war; the appearance of the chameleon represents war in the particular case. The change in character results from a change in the environment in which the chameleon finds itself. Its surroundings exist outside and beyond itself, beyond its control. Clausewitz is making the observation that the character of war will always be shaped by the context in which it takes place – the chameleon changes its colour, but it does so more or less involuntarily as a result of its environment. Yet, war is certainly 'more than' a chameleon in another sense.

The analogy does not properly convey the complex, interdependent relationship between, on the one hand, war and those engaged in war, and on the other, the context within which they wage war. As Clausewitz was aware, there is a continual, reciprocal relationship between the two. The popularly conceived chameleon simply reacts to its surroundings and adapts accordingly, but this is not the case with the social phenomenon of war. The context is, to a great extent, formed by the actions of individuals and groups and they then act within that context.

Clausewitz's analogy of a chameleon is the clearest expression of the concept of context in the trinity. As such, it embodies a whole range of insights found throughout Clausewitz's works. Paret neatly summarises this when he states that the three elements of war are 'not compromised or distorted by being placed in a particular historical context, the context allows us to see their essential nature as they act and react in a dynamic state rather than as abstractions'.[48] To conclude this section, we should return to Clausewitz's historical survey in Book 8, Chapter 3B.

45 *On War*, I, 1, § 28, p. 101.

46 Clausewitz's use of metaphor is discussed in Alan Beyerchen's article, 'Clausewitz, Nonlinearity and the Importance of Imagery'. Available online at: <http://www.clausewitz.com/readings/Beyerchen/BeyerschenNonlinearity2.pdf>, retrieved 20 April 2007.

47 Christopher Bassford, 'The Primacy of Policy and the "Trinity" in Clausewitz's Mature Thought', in Strachan and Herberg-Rothe, *Clausewitz*, p. 78.

48 Paret, *Understanding*, p. 23.

The crucial final passage is a concise expression of Clausewitz's broad theoretical framework, as embodied in the trinity, and thus enables us to clearly determine the place context assumes in it. Clausewitz states that:

> The aims a belligerent adopts, and the resources he employs, must be governed by the particular characteristics of his own position [secondary level]; but they will also conform to the spirit of the age and to its general character [context]. Finally, they must always be governed by the general conclusions to be drawn from the nature of war itself [primary level].[49]

There is thus considerable evidence of Clausewitz's reliance on the broad notion of context – or 'conditions' – in relation to war. Its presence in the trinity as a partially hidden, yet central theoretical device has also been demonstrated. However, if we are to reaffirm the place of context in our interpretation of the trinity, the concept requires deeper analysis in relation to recent scholarship and in the light of historical experience.

Explorations

Clausewitz certainly takes us quite far in understanding context, but a closer consideration reveals a number of potential problems with this slippery concept. Context, no matter how commonsensical its inclusion as a vital aspect of a theory of war may appear, is by no means a simple idea. In fact, it is so diffuse that fixing on any one simple description of its central attributes is extremely difficult. Therefore, the best way to proceed is to probe the concept from a number of angles, by asking different questions of it and considering its boundaries with other concepts.

Much of the best modern scholarship on war stresses the importance of context in a manner reminiscent of Clausewitz and, indeed, often in conscious emulation of him. Colin Gray, a leading strategic theorist and self-confessed Clausewitzian, has described context as 'the most important variable in understanding war'.[50] In his book, *Another Bloody Century*, he asserts that the only way to plausibly discuss future warfare is to embed the analysis in war's multiple contexts.[51] Indeed, this is a unifying theme in his numerous publications on strategy. In Rupert Smith's important book, *The Utility of Force*, his more overtly pedagogical thesis is built

49 *On War*, VIII, 3B, p. 718.

50 Colin S. Gray, 'How Has War Changed Since the End of the Cold War?', *Parameters*, Spring, 2005, p. 14.

51 Colin S. Gray, *Another Bloody Century: Future Warfare* (London: Weidenfeld & Nicolson, 2005).

entirely around the idea of contextual change.[52] That modern wars are fought predominantly 'amongst the people' in contrast to those of the era of industrial war, is a proposition based on the notion of fundamentally changed conditions. Lawrence Freedman's thesis outlined in his Adelphi Paper on *The Transformation of Strategic Affairs* similarly emphasises the impact changes in context – principally in political conditions – have on Western strategic choices in the modern world.[53]

The substantive content of these works is suffused with deep contextual knowledge, which only serves to enhance the integrity and worth of the analysis. Nevertheless, there is a hidden assumption that we understand just what context is. It is necessary to examine the concept itself (as opposed to its substantive manifestations) in greater detail. As with any analytical concept, we need to understand where it begins and ends in relation to other concepts. Context is particularly demanding of such definitional interrogation as it has become a greatly overused term, sapping it of meaning and explanative power.

Gray's discussion of strategic culture in his *Modern Strategy* offers us a useful starting point. Gray notes a dualism in definitions: 'context can be considered as something "out there", typically in concentric circles, meaning "that which surrounds". Alternatively … one can approach context as "that which weaves together" (from the Latin *contextere*, "to weave together")'.[54] The first perspective certainly corresponds to typical conceptions. The dictionary defines context as 'the circumstances that form the setting for an event, statement, or idea'.[55] The second of Gray's perspectives is less common but, as explored below, is central to a proper understanding.

The first perspective, of context as that which surrounds, logically entails that the concept only becomes meaningful when used in relation to a particular 'something' (be it event, statement or idea). Understanding the substance of context only becomes possible when considered in relation to a particular situation or thing. We can ask the question 'what is context?', however, one cannot plausibly ask in isolation, 'what is the context?' without clarifying what we want to know the context of. Thus, the first basic point is that context only has *substantive* meaning in relation to a particular subject or 'thing'.

Although this must be accepted, context is, to a certain extent, always 'out there' in that it represents the permanent setting in which events are played out. So, rather than the notion that context requires a subject, this perspective holds that subjects can never escape context. The dimensions that come to be described as the context of a particular event are permanently extant. Those dimensions

52 Rupert Smith, *The Utility of Force: The Art of War in the Modern World* (London: Penguin, 2005).

53 Lawrence Freedman, *The Transformation of Strategic Affairs*, Adelphi Paper 379 (London: Routledge, 2006).

54 Gray, *Modern Strategy*, p. 130.

55 Catherine Soanes (ed.), *The Oxford Dictionary, Thesaurus, and Wordpower Guide* (Oxford: Oxford University Press, 2001), p. 257.

only truly become 'the context' of something when that something happens. To describe context as entailing a vague realm existing independently from any particular event certainly puts strain on the concept defined simply as 'that which surrounds'. The things that come to be described as the substance of context could be described as context waiting to happen.

Context and time – change and continuity

How does context relate to time and change? When we consider different dimensions of context below, one point becomes clear: that the various substantive dimensions display different degrees of change or stasis over time. Context, as an operative concept in the theory of war, is permanent: no actor exists outside of context. Yet context in its substantive form has the potential for endless change over time. This is reflected in the unique circumstances in which actors find themselves and in the character of their own institutions, beliefs or available technologies. Context – in terms of its dimensions – can be understood as a spectrum allowing for gradual change (such as geography, aspects of culture or, we might add, human nature) to rapid or sudden change (such as technological advances or new political situations brought about, for instance, by revolution). As a holistic concept context never displays complete stasis: there is constant change in some of its dimensions regardless how gradual they may be. The idea of *change* is thus integral to the concept.

Related to this idea, there is an analytical problem here of distinguishing or making judgements on the rate of contextual change. Which aspects of contextual change we emphasise *depends on what we are attempting to explain.* In a variation on the bureaucratic axiom that 'where you stand depends on where you sit', how we evaluate a particular contextual dimension depends on what it is we are attempting to understand. It is this issue which makes abstract discussion of context so difficult. Judgements on the extent, rate and form of contextual change are relative and subjective. Similarly, recognition of relative levels of change in context is not to pronounce on the comparative difficulty or ease of strategy – that is dependent on factors peculiar to any particular war. More stable conditions allow protagonists to become better acquainted with their environment, their opponents and themselves. Yet, even then, marginally different circumstances, varying objectives or the unpredictable moves of the enemy can render irrelevant what comfort stable contexts may provide.

Recognition of significant contextual change in relation to war is a subject strongly associated with so-called 'military revolutions'.[56] Historians have identified what they perceive to be major changes in the character and conduct

56 See Williamson Murray and MacGregor Knox, *Dynamics of Military Revolution, 1300–2050* (Cambridge: Cambridge University Press, 2008) and Williamson Murray, 'Thinking About Revolutions in Military Affairs', *Joint Forces Quarterly*, Summer 1997, pp. 69–76.

of war resulting from or influenced by transformations, 'earthquakes' or systemic changes in politics, society, technology and administration.[57] The debates that surround exactly when, where and why such revolutions have taken place is not our principal concern here, but the idea sheds light on the nature and potential impact of context on war.

By way of example, one of the most important and far-reaching military revolutions accompanied the formation of centralised nation-states in the early modern period, roughly dated to the period between 1500 and 1800. The financial burdens of war resulting from new technologies (such as gunpowder weapons, elaborate fortifications and gunned ships) meant only centralised states with bureaucratic machineries and sophisticated money markets could generate the economic wealth through taxation and deficit financing to maintain effective armed forces and the socio-political infrastructures to organise them.[58] Resource extraction thus became a prerequisite of power, while wealth became increasingly dependent on military effectiveness in wars supporting mercantilist trading empires.[59] This period also saw the emergence of recognisably modern, regular, professional, larger armies – from the Spanish *tercios*, through Gustavus Adolphus's Swedish national army, to Louis XIV's royal army[60] – which replaced the localised feudal levies of the middle ages and the unpredictable *condottieri* companies of the Renaissance.[61] Princely power was consolidated, medieval 'private war' declined and the modern state with its own professional standing armed forces, dedicated to the promotion of its national interest, took shape. The path to absolutist, national and 'total' wars lay open.

Of course, although he didn't express it in such terms, Clausewitz essentially believed he had lived through and witnessed a profound military revolution, sparked by the enormous social and political upheavals associated with the French Revolution. Few today would disagree with the basis of his observation.

Context: weaving together

An important point that might not be clear from what has been written so far is that context should not be seen as wholly external to the actors in war. The concept runs through and within the social entities under observation. A crucial aspect of context relates to a given actor's own position: its form, character, norms, culture and ideology, as well as those of its opponents, allies and other relevant groups. Context should not be conceived as a wholly exogenous concept. It will be recalled that Clausewitz employed the analogy of the chameleon in conveying the idea of

57 Murray and Knox, *Military Revolution*, pp. 6–7.

58 Gat, *War in Human Civilisation*, pp. 484–9.

59 Gat, *War in Human Civilisation*, p. 472 and pp. 489–96. The point is also made by Howard in *War in European History*, p. 48.

60 See Howard, *War in European History*, pp. 32–4, pp. 57–60 and pp. 63–6.

61 Howard, *War in European History*, pp. 20–37.

how context impacts on war and it was noted that this was a useful yet insufficient metaphor; the limitation implied by Clausewitz's qualifier 'more than'. Likewise, the analysis so far suffers from a crucial deficit.

Although it is accurate to treat context as something that surrounds war, this might convey the impression that context is something that exists as an entirely independent realm, like an omniscient and omnipotent God directing the universe, whereby human actions and events are essentially preordained. In fact, context is more analogous to the Homeric Greek pantheon, composed of many varying dimensions, extremely capricious and intimately and reciprocally interwoven with human affairs. If we are to develop a more robust conception of context then we must understand the ways in which context not only surrounds, but as Gray stresses, 'weaves together'.[62]

The basic point is that humans and the societies, nations and institutions of which they are part, constitute an important element of the conditions of war. War is a social and political activity, and thus conditions relating to societies, institutions and cultural forms represent some of the most important factors impinging on its conduct, and these cannot be understood separate from the humans that comprise them. It is in this way that the trends and forms of human social conduct produce and reproduce the changing contexts within which the same actors conduct wars. The relationship between war's context and warfare itself is thus one of reciprocal and dynamic interaction. Due to this interaction, there is an extent to which, in some cases, powerful actors are able to shape their own contexts to varying degrees.

As this idea of 'weaving together' is perhaps somewhat abstract and difficult to grasp, it is useful to consider the issue in relation to the subject of strategic culture. The concept refers to the internalised ideas, patterns, prejudices and habits of behaviour that unique security communities display. It emerges from socially transmitted ideas, attitudes, habits and preferred methods of operation, which in turn derive from collective perceptions and interpretations of historical experiences, as well as feeding on all the dimensions of context discussed below. Strategic culture approximates to a form of 'guide to action' and can be particularly influential in times of crisis, when cultural instinct often kicks in.[63] As Black states, it is crucial to understand the 'plasticity or changeability' of the term, and that there may be multiple cultures within individual nations or groups.[64] Nor does culture imply a rigid causality between ideas and behaviour, tradition and action. Groups can display 'counter-cultural' behaviour, but in the *longue durée* such instances tend to confirm the overriding attitudinal environment as they tend to revert to type. For

62 Gray, *Modern Strategy*, p. 130.

63 Gray, *Modern Strategy*, pp. 129–51.

64 Jeremy Black, 'War and International Relations: a Military Historical Perspective on Force and Legitimacy', in David Armstrong, Theo Farrell, and Bice Maiguashca (eds), *Force and Legitimacy in World Politics* (Cambridge: Cambridge University Press, 2005), p. 136.

instance, the massive British continental commitment during World War I did not fundamentally alter its traditional maritime orientation.[65]

Generally, when groups go to war, they do so possessed of unique sets of assumptions, predispositions and habits with regard to the use of force. Decisions will at least be influenced by prevailing cultural preconceptions and norms: 'the human hosts of strategic culture are inalienably part of their own strategic context'.[66] This culture may be so deeply ingrained that we can begin to talk of a nation's 'way of war' – continuities in the behaviour of certain strategic cultures. This represents an important contextual feature for any nation or armed group; one that is inextricably part of its own unique history and institutions. It is one notable way in which groups engaged in war become, to an extent, an element of their own context.

Context: important, but not decisive

There is another essential point regarding the concept of context that must be clarified. Clausewitz's emphasis on contextual understanding is, at heart, simply a recognition of its considerable importance as a permanent factor influencing war. This line of argument is not equivalent to suggesting that contextual factors are always the major determinants of success or failure in war. It does not entail subscribing to any overarching deterministic causality: agency still matters. It is perfectly possible for commanders to achieve the defeat of the enemy with little or no comprehension of wider contextual matters. Ultimately, wars will be decided by events on the ground, decisions taken and, as Clausewitz stressed, by pure chance or luck. As Gray notes, 'the mystery of combat performance is by no means entirely the product of contextual factors but rather is driven in part by influences integral to war itself'.[67]

For the commander, contextual knowledge is no silver bullet. For the historian, context cannot explain all. Too much emphasis on context 'removes the sense of uncertainty in which contemporaries made choices'.[68] For the historian, hindsight is something of a poisoned chalice to the extent that it can encourage judgements of inevitability based on analyses of prevailing conditions. Of course, it is perfectly possible to misdiagnose context; indeed, contextual comprehension is notoriously difficult and potentially overwhelming,[69] and perhaps in such cases we may conclude it would have been better not to try (for the commander, leading to mistaken actions, for the historian, mistaken analyses). Nevertheless, as a general rule, careful consideration of context is to be encouraged.

---

65 Gray, *Modern Strategy*, p. 139.

66 Gray, *Modern Strategy*, p. 129.

67 Gray, *War, Peace and International Relations*, p. 35.

68 Black, *Global Military History*, p. 3.

69 Gray, *War, Peace and International Relations*, p. 10.

Context can be viewed as setting the broad parameters within which success in war is achieved, or that may equate to what Barry Watts usefully terms 'option sets in possibility space'.[70] For instance, states do not have much say as to where they are located, politicians must act within the geopolitical realities of the moment and commanders have to use the armies they have available to them. Such facts are the historically determined and largely inescapable conditions within which war is waged. A war will necessarily have to be considerably protracted for serious contextual change to make itself felt, beyond mere circumstantial alteration. Through certain measures humans can attempt to solve or mitigate the problems posed by conditions, but they cannot escape context itself.

For instance, it is axiomatic in modern counterinsurgency doctrine that winning the allegiance of the people to the government is the sure route to success and many practical implications flow from that premise. Yet, that assumption is heavily context dependent in that it largely relates to a specifically Western liberal-democratic approach. Historically, the alternative for many tyrannical regimes has been a far more direct military, forceful and repressive approach along the lines of 'kill them all, let God sort them out'.[71] John Nagl notes the approach adopted by King William II in suppressing a Welsh revolt in the late eleventh century in which he 'intended to abolish and utterly destroy all of the people until there should be alive not so much as a dog'.[72] Unconstrained by the standards expected of liberal regimes or electoral politics, many autocrats have suppressed uprisings with calculated, audacious efficiency, if not wholesale slaughter. When liberal states have resorted to elements of such an approach (such as torture, massacres or forced resettlement), this has usually led to local or domestic political turmoil, as highlighted by the French experience in Algeria during the 1950s and 60s. In this light, the context of democratic domestic politics represents a crucial determinant that makes the minimum use of force and winning the support of the population more expedient, if still hugely difficult.

From the commander's perspective, a wilful disregard of context is a likely cause of strategic failure. Conditions may come close to dooming an actor strategically before the act of war itself. But, again, the result is not entirely determined, even if it may be *probable*. The impact of contingent factors pervade the whole course of war and make prediction impossible. At a minimum, contextual knowledge may enable strategists to anticipate likely enemy weaknesses or avoid potentially counter-productive actions. At best, shrewd exploitation of contextual knowledge may allow the commander to threaten force in a way that causes the enemy's capitulation before a shot has been fired.

---

70 Barry Watts, *Clausewitzian Friction and Future War* (Washington: National Defense University, 1996), p. 54.

71 Charles Townshend, 'Counter-insurgency', in Richard Holmes (ed.), *The Oxford Companion to Military History* (Oxford: Oxford University Press, 2001), p. 231.

72 John A. Nagl, *Learning to Eat Soup with a Knife: Counterinsurgency Lessons from Malaya and Vietnam* (Chicago: University of Chicago Press, 2005), p. 26.

Context sets the conditions in which certain strategies and tactics are likely to be effective. This largely explains why the opponents of Napoleon were unable to defeat the Emperor's forces until they adapted sufficiently to keep pace with the new form of war generated by radically altered social, political and, ultimately, military realities.

From the historian's perspective, a wilful disregard of context is a likely cause of parochial explanations. Yet we should not be over-impressed by what context can explain. Outcomes may have been as much the result of great generalship, chance happenings or random decisions taken on the day. Nevertheless, at a minimum, knowledge of context enables the historian to appreciate why belligerents are fighting, why they are employing various means as they are, why the public is reacting as it is and so on. At most, context may constitute the most important factor in the explanation of a strategic outcome. Context provides greater meaning to otherwise seemingly arbitrary events and enables deeper causal analysis. Historians are entitled to describe events as they happened, but they will make little sense if entirely detached from causative historical context.

Dimensions of context

Up to this point our discussion of context has taken place at a somewhat abstract level in order to determine some of its most important aspects. Yet, we need to develop a clearer idea of how context is manifested in reality. This section thus outlines those fundamental aspects of context that have served to shape the character of war-making peoples throughout the course of history. Yet, just how far can we go in theory on this subject? Given the inherent changeability of context, there is a limit to what can be stated with universal applicability other than to outline its most salient features.

Linked to this, it must be noted why only prominent dimensions will be considered here. Of course, it would be possible to widen our scope to incorporate all the multifarious aspects of context in their infinite variety, but such an undertaking is beyond the aims of this book. As Gray comments, 'One can offer too big a big picture'.[73] As Knox notes, we – and in particular those responsible for making strategy – must narrow our focus to some extent because 'too much complexity makes the mind seize'.[74] Here we are interested only in the fundamental dimensions of context and those that relate, demonstrably and consistently to war.

73 Gray, *Bloody Century*, p. 131.

74 MacGregor Knox, 'Conclusion: Continuity and Revolution in the Making of Strategy', in Williamson Murray, MacGregor Knox, and Alvin Bernstein (eds), *The Making of Strategy: Rulers, States and War* (Cambridge: Cambridge University Press, 2007), p. 645.

Historical context

Context makes little sense without understanding its connection to history or historical chronology. From one perspective, history provides context with its form: social relations do not exist in isolation from time; they are crucially affected by prior events and in our analysis of the 'here and now' the past cannot be ignored. All actors cannot but be affected by what has gone before them, either in the way history shapes material realities or through its impact on social and cultural forms. Understandings of what happened in the past powerfully shape the ideas and behaviour of those in the present. History often influences events in the most subtle ways as it feeds into the assumptions and beliefs of those acting in the present, but where the precise impact is difficult to fathom. It can inform perceptions about other groups and indeed, one's own group, based on past experiences or historical myth.

Examples of history shaping the causes and course of war are manifold. For instance, there can be little doubt that the spate of revolutionary wars in the third quarter of the twentieth century were powerfully influenced by the example of Mao's earlier successful guerrilla campaign in China.[75] History, often manipulated by unscrupulous leaders, can sometimes lend disproportionate importance to otherwise barren territory, inflame hatreds where peaceful coexistence once reigned or prevent resolution of a political problem where common ground might otherwise exist. Regarding the latter point, if between two peoples there exists a history of conflict and betrayal, peaceful solutions to current wars may prove even harder to achieve.

History also feeds into the institutional memory, doctrines and plans of war-making institutions.[76] The Napoleonic campaigns served as a major source of operational lessons for most European militaries up to World War I as the understanding of them was filtered through various interpretations.[77] This was particularly true in France where the offensive spirit of Napoleonic warfare was emphasised even as technological change appeared to suggest the primacy of the defensive. Earlier, in Clausewitz's Prussia, memory of past successes under Frederick the Great was one reason out-dated methods were retained: in this way, sometimes history, as context, can retard adaptation to changing conditions.

History weaves together political, social and cultural threads over time, and it represents a combination of traditions, experience and inherited beliefs. In particular, past wars may be considered the most potent aspects of history. If long, involving great sacrifices or marking important turning points in a group's history, they can powerfully shape attitudes and behaviour down the years (even ones

75 Brian Crozier, 'The Strategic Uses of Revolutionary War', in Alistair Buchan (ed.), *Problems of Modern Strategy* (London: Chato and Windus, 1970), pp. 198–219.

76 Hew Strachan, *European Armies and the Conduct of War* (Kings Lynn: Routledge, 2006).

77 Gray, *War, Peace and International Relations*, p. 34.

fought hundreds of years ago, whether victory or defeat). For instance, the conflict in Kosovo between Serbs and Albanians during the 1990s was pervaded by the memory of the Serbian defeat in the 1389 Battle of Kosovo Polje. Tim Judah has even remarked that, in Kosovo, 'history is war by other means'.[78] In this way, history becomes inextricably intertwined with political realities today.

Political context

This dimension encompasses a wide variety of elements, foremost amongst which are the nature of the political tensions between groups, the distribution of power in the system and the conditions and relationships within and beyond states leading up to and during war. Here we are concerned with prevailing power dynamics, be they local, domestic, regional or international. The political context is fundamentally characterised by continual change and is intertwined with other contexts such as economic power, technological sophistication or dominant values, all of which feed into power constellations surrounding and running through war. Morphing political conditions over time generate the immediate conditions which lead to conflict between groups and serve as basic determinants of war's character throughout and into the subsequent peace. As Gray notes, 'The political context is the source of, and provides the meaning for, war and its conduct in warfare'.[79] The political 'womb in which war develops'[80] can be comprised of a complex mixture of internal and external factors which coalesce to cause a state of confrontation between groups.

So, for instance, in his analysis of the wars of his age, it was changed political conditions which Clausewitz stressed as the fundamental determinants of their cause and character, both with respect to the internal permutations emanating from the consequences of the French Revolution, but also regarding the impact of these changes on the wider political order of Europe. The ideals embodied in the new republican French state represented a serious ideological threat to the cherished traditions and social patterns of the conservative states of Europe, while – as Edmund Burke powerfully argued[81] – the subsequent empire building of Napoleon threatened the unwritten, yet powerful stabilising norm of the balance of power on which relations between *ancien régime* states had been based for at least a century or more.[82]

---

78 Tim Judah, *Kosovo: War and Revenge* (New Haven: Yale University Press, 2000), pp. 1–32.

79 Colin S. Gray, *Recognizing and Understanding Revolutionary Change in Warfare: The Sovereignty of Context* (Carlisle, PA: Strategic Studies Institute, 2006), p. 15.

80 *On War*, II, 3, p. 173.

81 Chris Brown, Terry Nardin, and Nicholas Rengger (eds), *International Relations in Political Thought: Texts from the Ancient Greeks to the First World War* (Cambridge: Cambridge University Press, 2003), p. 254.

82 Friedrich von Gentz, 'The True Concept of a Balance of Power', in Brown et al., *International Relations in Political Thought*, p. 306.

Also, domestic political conditions constitute powerful forces on war. Their impact is particularly apparent in relation to the wars conducted by modern liberal democratic states. With their free media, civil society and frequent electoral cycles, leaders in wartime are faced with profound political and military conundrums. This is particularly noticeable in Western 'wars of choice' due to the difficulty of explaining the necessity of intervention, justifying mounting casualties and resisting pressure for precipitate 'exit plans' when often the overriding strategic requirement is a will to stay the course and demonstrate to local populations that they will continue to provide security.[83]

Cultural context

In an essay on 'The Germans and the French' written in 1807, Clausewitz notes that 'the character of a nation, even insofar as its original, fragile nature may be affected by changing customs, is not as easily reshaped as philosophers and moralists seem to think'.[84] This idea is presented in a discussion about the potential impact of national character on war and conveys the importance he attached to socio-cultural conditions. Written soon after the disaster of 1806, he notes that, 'I cannot conceal it from myself how far the German national character has contributed to our present condition'.[85]

Prevailing social forms, cultural norms, attitudes, traditions, ideologies, religious beliefs and values are all crucial factors influencing the complexion of war. Whereas traditional approaches tend to focus on material features, Jeremy Black stresses that it is 'far from clear that variations … in "cultural" factors, and related norms, should play a smaller role in the history of war than weaponry'.[86] As noted, these elements are strongly connected to history in the way such forces are moulded over time by events, interaction with other communities and the impact of changing material conditions.

Cultural factors may constitute crucial determinants of behaviour, shaping decision makers perceptions regarding strategic alternatives. If, for instance, we can speak of a modern Western 'way of war', it is strongly shaped by prevalent socio-cultural forces. Black has identified a number of prominent trends in this respect. He points to a 'Revolution in Attitudes towards the Military' influenced by the spread of democratisation, individualism, hedonism, changed gender relationships and a burgeoning human rights discourse,[87] which have contributed

---

83 Paul Cornish, 'The United States and Counterinsurgency: "Political First, Political Last, Political Always"', *International Affairs*, Vol. 85, No. 1, January 2009, p. 62.

84 Carl von Clausewitz, 'The Germans and the French', in Clausewitz, *Writings*, p. 252.

85 Clausewitz, 'The Germans and the French', p. 252.

86 Jeremy Black, *War Since 1945* (London: Reaktion Books, 2004), p. 13.

87 Black notes that this has resulted primarily from social shifts, combined with ideological and cultural pressures. Jeremy Black, *War and the New Disorder in the 21st Century* (London: Continuum, 2004), p. 18.

to a decline of civic militarism, societal debellicisation and a general reliance on professionalised forces.[88] Western publics thus demand that collateral damage and both civilian and military casualties are kept to a minimum, particularly in wars of choice where losses are deemed unnecessary.[89] This in turn has contributed to a reliance in Western military operations on precision air-power and technological virtuosity as well as force protection. This is of course just one general example. It is important to keep in mind the vast possibilities encompassed in this dimension, be it manifested in the culture of glory and imperial ambition of Rome, the religious fanaticism of the European middle ages or the 'secular religions' of the nineteenth and twentieth centuries.

Legal and ethical context

Clausewitz ostensibly downplays the importance of the law and ethical considerations in war; a position consistent with many classical realist thinkers. Famously, he states that, 'Attached to force are certain self-imposed, imperceptible limitations hardly worth mentioning, known as international law and custom, but they scarcely weaken it'. Yet, we must recognise that he makes this point in the context of his abstract discussion of absolute war, where he demonstrates that, *logically*, such factors do little to limit the inherent escalatory dynamic of war. Nevertheless, Clausewitz's overall attitude is certainly one that downplays the role of law or morality and which views the use of force in the prevailing *realpolitik* conception. In true Machiavellian style, Clausewitz remarks that 'mistakes which come from kindness are the very worst'[90] and 'Woe to the government, which relying on half-hearted politics and a shackled military policy, meets a foe who, like the untamed elements, knows no other law than his own power!'[91]

However, a closer analysis reveals a more sympathetic and inclusive conception. It is apparent that Clausewitz would at least have been aware of the potential influence of such factors, as even the extreme exponent of *realpolitik*, Machiavelli recognised before him. As Howard notes of Clausewitz, 'He knew very well ... that the conduct of war was subject to considerably greater and more perceptible limitations in his own times than it had been in the days of, say, Genghis Kahn'.[92] Also, his dismissive tone is balanced somewhat by the importance he attaches to social norms: that armies in the eighteenth century did not generally plunder and lay waste to the enemy's land largely owed, he suggests, to 'the spirit of the

88 Black, *Disorder*, pp. 13–25.

89 McInnes, *Spectator-Sport Warfare*, p. 72.

90 *On War*, I, 1, § 3, p. 84.

91 *On War*, III, 16, p. 257.

92 Michael Howard, 'Constraints on Warfare', in Michael Howard, George Andreopoulos, and Mark Shulman (eds), *The Laws of War: Constraints on Warfare in the Western World* (New York: Yale University Press, 1994), p. 2.

times'.[93] While most limitations in war he attributed to practical considerations, the balance of power or the more effective use of force, Clausewitz did not entirely reject the potential influence of law.

It is important to note that the laws of war and ethical traditions have exerted differing levels of influence throughout history. From the customs of primitive tribes to Chivalric codes and the Geneva Conventions, societies have sought to establish when war is justified and what is deemed acceptable in war. In the last century, the laws have extended to cover a great many issues. For much of history, the laws of war have consisted largely of unwritten social understandings, while powerful ethical traditions – such as that of the Christian 'Just War' and its secular juristic descendants – have shaped public discourse and the decisions of political and military leaders.[94]

The last 150 years has seen the increasing codification of the 'laws of war', while ethical thinking has tended to converge with juristic conceptions.[95] The importance placed on these constraints by states is testified by the legions of lawyers authorising attacks on individual targets in modern military operations. These are all crucial developments that testify to the remarkable developments in this area, even if, as throughout history, the laws of war are observed more in breach than observance. The trinitarian framework must be capable of incorporating these factors even if Clausewitz did emphasise, perhaps rightly, the limits of law and moral concern in an activity so pervaded by necessity.

Economic context

It is easy to forget that, 'Wars, after all, must be paid for'[96] and as Strachan notes, 'it remains helpful to consider the economic constraints on military policy as a distinct pressure on the development of war's conduct'.[97] Troops have to be paid, fed and supplied, weapons procured and so on: as Gray reminds us, this dimension is 'fundamental, enduring, comprehensive, and inescapable'.[98] Throughout history, wars have been shaped by what is and what is not possible economically. From a broader perspective, economic wealth has constituted a major factor in determining

93 *On War*, VIII, 3B, p. 714.

94 Nicholas Rengger, 'On the Just War Tradition in the Twenty-First Century', *International Affairs*, Vol. 78, No. 2, 2002, pp. 353.

95 Nicholas Rengger, 'The Judgement of War: On the Idea of Legitimate Force in World Politics', in David Armstrong, Theo Farrell, and Bice Maiguashca (eds), *Force and Legitimacy in World Politics* (Cambridge: Cambridge University Press, 2005), pp. 150–52.

96 Mark Grimsley, 'Surviving Military Revolution: The U.S Civil War', in Knox and Murray, *Military Revolution*, p. 84.

97 Strachan, *European Armies*, p. 6.

98 Gray, *Modern Strategy*, p. 32.

military superiority and the relative power of nations.[99] In modern times, it has been argued that economic factors attained an almost strategic lead role in the two World Wars.[100] Military operations are often fought to secure resource rich territory to further fuel war. For instance, Hitler's annexation of the Sudetenland in 1938 was partly undertaken to secure additional resources to boost Germany's faltering rearmament programme.[101]

At least since the demise of mercantilism in the eighteenth century, prominent commentators have held that the context of increasing global economic interdependence will foster peace, given the increasing costs of going to war.[102] This hope was famously dashed in 1914 and ever-increasing globalisation has failed to curb the outbreak of war in recent years, perhaps even exacerbating tensions leading to conflict.[103] Niall Ferguson has argued that the wars of the twentieth century were propelled forward by economic volatility, which intensified socio-political tensions, promoted instability in weakening empires, undermined new democracies and heightened racial antipathies.[104] Studies of contemporary conflict in the developing world have drawn attention to the prominent role of 'greed', struggles to control natural resources and economic inequalities between culturally-defined groups in fuelling organised violence. In a similar vein, many prognostications of future conflict point to the potential for conflicts caused by resource scarcity and social ruptures caused by economic tensions.

It is common to note that Clausewitz paid little heed to the economic dimension of war. He barely discussed Napoleon's 'Continental System' and Knox notes how he was surprisingly 'mute on the revolution in the mills and iron-foundries that had kept Britain fighting until triumph at Waterloo',[105] yet it would be mistaken to claim that he did not understand its general importance.[106] Clausewitz notes how, at the end of the seventeenth century, the power of states 'lay entirely in their treasuries' as money payments, supported by a more efficient administrative machinery, allowed central governments to recruit large standing armies.[107] Through the eighteenth century, the means which states had to wage war came to be based largely on the available 'money in their coffers'. Thus, Clausewitz

99 Azar Gat, *War in Human Civilisation* (Oxford: Oxford University Press, 2006), pp. 489–96; Paul Kennedy, *The Rise and Fall of the Great Powers* (London: Fontana Press, 1989).

100 Knox, 'Conclusion', p. 640.

101 Wilhelm Deist, 'The Road to Ideological War: Germany, 1918–1945', in Murray et al., *Makers of Strategy*, p. 383.

102 Quoted in Gat, *War and Civilisation*, p. 510.

103 Mary Kaldor, *New and Old Wars: Organized Violence in a Global Era* (Cambridge: Polity, 2002), pp. 69–70.

104 Niall Ferguson, *The War of the World: History's Age of Hatred* (London: Penguin, 2006), 'Introduction', pp. xxxiii–lxxi.

105 Knox, 'Conclusion', p. 639.

106 Many of his historical studies refer to economic and financial factors.

107 *On War*, VIII, 3B, pp. 711–12.

clearly recognised that the financial context was a major factor shaping the wars of the age.[108] Many of his historical studies refer to the importance of economic and financial matters.

Although the economic dimension of war is important – and has perhaps been close to decisive in some modern wars – it is perhaps useful to remember Machiavelli's advice that it is 'imprudent to accept the common maxim that riches are the sinews of war'.[109] Economic factors should be viewed as strategic enablers – wars still have to be fought and won at the operational and tactical levels,[110] and as the Americans learnt in Vietnam, massive economic superiority is no guarantee of success. Similarly, Richard Overy has dismissed the popular thesis that Gross Domestic Product won World War II.[111]

Economic factors, like technology, are best considered as crucial material contextual features, but which do not warrant inclusion as part of war's inherent nature, as some commentators have proposed.[112] As Stephen Biddle argues, material preponderance has been exaggerated and states' relative economic, demographic or industrial strengths are poor indicators of military power; they only matter if they can be effectively exploited through modern doctrines of tactical force employment.[113] Similarly, Paret remarks that Clausewitz was too knowledgeable to equate mere wealth with military strength.[114] Of course, perhaps the principal reason economic power is crucial in war is so that political entities have the ability to invest in, produce or buy the myriad material implements of war. After all, as Machiavelli stated, 'war is made with steel and not with gold'.[115] This brings us to the technological context of war.

Technological context

Another frequent criticism levelled at Clausewitz's conception of the trinity is that it fails to sufficiently address the importance of technology as an integral element of war. This was probably due in large part to the fact that the rate of technological change had been relatively gradual in the period prior to and around the time he wrote. The advances in technology since the early nineteenth century have indeed

108 *On War*, VIII, 3B, p. 713.

109 Skinner, *Machiavelli*, p. 85.

110 Gray, *Modern Strategy*, pp. 32–3.

111 Richard Overy, *Why the Allies Won* (London: Pimlico, 1996), pp. 316–17.

112 Ian Roxborough, 'Clausewitz and the Sociology of War', *The British Journal of Sociology*, Vol. 45, No. 4, December, 1994, pp. 619–36. Michael I. Handel, 'Clausewitz in the Age of Technology', in Michael I. Handel (ed.), *Clausewitz and Modern Strategy* (Abingdon: Routledge, 2005), pp. 51–92.

113 Stephen Biddle, *Military Power: Explaining Victory and Defeat in Modern Battle* (Princeton: Princeton University Press, 2004), pp. 3–11 and pp. 190–96.

114 Paret, *Clausewitz*, p. 209.

115 Machiavelli quoted in Quentin Skinner, *Machiavelli: A Very Short Introduction* (Oxford: Oxford University Press, 2000), p. 86.

been far-reaching.[116] Yet, Clausewitz was right not to view technology as part of war's intrinsic nature, but rather as a contextual dimension. As he states, 'It is clear that weapons and equipment are not essential to the concept of fighting, since even wrestling is fighting of a kind'.[117] While the relationship of technology to all dimensions of war is important, we must maintain perspective. Technological superiority alone has rarely assured victory and, moreover, in the longer term, 'Every new device and mode of war carries the virus of its own technical, tactical, operational, strategic, or political negation'.[118]

The technological dimension of context is a vast subject covering many subsidiary elements, including weapons, communications, transport, intelligence collecting equipment or even food preservation.[119] That technology has, since the beginning of human history, constantly shaped and reshaped the waging of war in important ways is an unremarkable observation. Since the scientific and industrial revolutions, the pace of technological change has accelerated and the potential material means available to humans have proliferated – there simply is not room here to describe such changes in detail. More recently, as will be discussed further in Chapter 5, technology has been the driving force behind a so-called 'Revolution in Military Affairs', whereby precision weapons, numerous high-tech sensors and information technologies have been exploited to enhance military effectiveness. As with the introduction of revolutionary new technologies in war throughout history, no doubt the character or 'face' of war today has been transformed, yet, as history also suggests, such technology will not provide independent solutions to strategic problems as some commentators predict.

Geographical context

Finally, within the geographical dimension of context we include, first, such factors as the location, size and features of the territory of political entities and their geopolitical relation to one another. Second, it encompasses the various spatial dimensions within which war is fought, which have radically expanded over the last two centuries – initially being confined to either land or sea, commanders now have to cope with war in the air, space and even cyber space.[120] These developments – the consequences of technological change – have undoubtedly rendered modern warfare more geographically diverse and complex. Clausewitz observes that 'a commander must submit his work to a partner, space … which because of the constant movement and change to which he is subject he can never really come to know'.[121] Third, the physical environment, including terrain, topography and

116 Knox, 'Conclusion', p. 639.
117 *On War*, II, 1, p. 145.
118 Gray, 'Why is Strategy so Difficult?', *Joint Force Quarterly*, Summer 1999, p. 9.
119 Black, *Disorder*, p. 9.
120 Gray, *Modern Strategy*, p. 40.
121 *On War*, I, 3, p. 127.

the weather are all geographical factors that greatly impinge on the character and course of war. Clausewitz notes that 'combat uninfluenced by its surroundings and the nature of the ground is hardly conceivable'.[122]

The simple fact of a nation's location can have crucial strategic and geopolitical implications. As Gray notes, 'Geography is destiny'.[123] A state's location can greatly shape its strategic culture or tactical doctrine: consider the different imperatives faced by a continental state such as Germany and an island nation like Great Britain.[124] The size of a nation may have crucial implications. For example, the vast Russian hinterland enabled resistance against both Napoleon and Hitler by 'trading space for time in a protracted defence'.[125]

Weather, terrain and other elements of the environment can be 'capricious, disadvantageous, fortuitous, formidable, opportune, problematic, and risk-inducing'.[126] Of course, for most of history the nature of the seasons, the need for soldiers to help with the harvest, combined with the constraints of existing transportation, restricted campaigns to a few months of the year. Yet the elements have had a more direct impact on war. Indeed, weather extremes have been responsible for some of history's most notable events in war: Napoleon's horrific crossing of the thawing Berezina River in 1812; General Burnside's abortive 'Mud March' of 1863; and the Russian winter which essentially saved the Soviet Union in 1941.[127]

Such environmental factors are difficult to predict except in general terms; history certainly suggests commanders should expect the unexpected.[128] Some aspects of the topographical context are naturally occurring, such as rivers and mountains, while others might be man-made, in particular in the form of fortifications, barriers and security walls of various descriptions. Technology has of course been used to overcome some of the more limiting effects of war's geographical dimensions, whether through specialised equipment, long-range weapons systems or advanced communications. Nevertheless, geographical dimensions remain an ever-present influence: 'time and distance and weather still exercise enormous influence on the strategic options and capabilities of states'.[129] Clausewitz notes that commanders require great imagination to be able to comprehend the manifold dimensions of their spatial and geographical context.[130]

122 *On War*, II, 2, p. 165.

123 Gray, *Bloody Century*, p. 75.

124 Williamson Murray and Mark Grimsley, 'Introduction: On Strategy', in Williamson Murray, MacGregor Knox, and Alvin Bernstein (eds), *The Making of Strategy: Rulers, States and War* (Cambridge: Cambridge University Press, 2007), p. 8.

125 Murray and Grimsley, 'Introduction', p. 9.

126 Harold A. Winters, *Battling the Elements: Weather and Terrain in the Conduct of War* (Baltimore: John Hopkins University Press, 1998), p. 266.

127 Overy, *Allies*, p. 314.

128 Winters, *Elements*, p. 268.

129 Murray and Grimsley, 'Introduction', p. 9.

130 Clausewitz discusses this in relation to what he calls the 'sense of locality'. *On War*, I, 3, pp. 127–8.

*

As MacGregor Knox has highlighted, human history in the last few centuries has undergone a profound transition from the 'cyclical quasistagnation of the agrarian age to the permanent revolution of science, technology, and industry'.[131] These momentous changes have certainly transformed the conditions within which war is conducted and at a seemingly accelerating rate. In this chapter we have sought to provide a general theoretical overview of the nature, dynamics and forms that context can take. The processes are too vast and endlessly changeable to allow for any detailed description; it is in the nature of the concept that it allows of concrete analysis only in relation to the particular. Rather, we have been concerned with grasping its importance at a theoretical level as an overarching, shaping influence on the forms that war can take, but which leave its essential nature relatively undisturbed.

These explorations have important implications for our understanding of the trinity. In order to fully understand the operation and meaning of its different elements and levels we need to firmly anchor the concept of context to the framework. The idea of context is already present in Clausewitz's trinity, but in a somewhat diffuse, unclear and hidden form. This has caused much misunderstanding and a failure to appreciate the in-built idea of *change* within the parameters of his theory. A number of commentators have noted the inherent flexibility or elasticity of the trinity and this is best understood when we bring together Clausewitz's conception of the essential nature of war with that of context.

It is the play of variable conditions on actors in war that explains the infinitely changeable manifestation of the three primary elements of war's nature at the secondary, subjective level. As Gray notes, 'the social, political, and even ethical, contexts are ever in flux. But … that fact poses no real difficulty for Clausewitz's theory of war'.[132] By bringing context firmly 'back in', we can better unleash the explanative potential of the trinity. Context is best conceived as forming a tertiary level in the framework.

Context, as an operative element of the trinity, should be understood as an ever-present multifaceted web comprised of a number of differing interconnected dimensions that interact over time. Context surrounds war but is also reciprocally and mutually dependent on the behaviour of actors and the development of events. The dimensions of context change over time in myriad ways, sometimes with a profound impact on the conduct of war that may be said to amount to a 'military revolution'. War itself is a major factor working to shape context over time.[133]

That Clausewitz emphasised the importance of unique conditions in war does not mean that he believed no generalisable insights could be extracted from

131 Knox, 'Conclusion', p. 615.

132 Colin S. Gray, 'Clausewitz, History, and the Future Strategic World', *The Occasional,* No. 47, June 2003, p. 6.

133 Gray, *Bloody Century*, pp. 56–7.

history. It did, however, certainly play a moderating role in the formulation of his theory. His recognition of contextual particularism protected his theory from limited and one-sided conclusions and led him to establish a high threshold for those aspects of the nature of war that can be held to be truly universal. Context, properly understood, is not part of the nature of war itself, but that nature, as expressed in the trinity, cannot be understood without reference to context, as it helps us explain why war's nature manifests itself in divergent ways throughout history. Or as Gray has put it, although change 'in the distribution of power, in the political culture of key polities, in the technological and economic contexts, and so forth – matters, it does not matter to the degree that it can transform the nature of the subject'.[134] As Clausewitz stated, 'we must face the fact that war and its forms results from ideas, emotions, and conditions prevailing at the time'.[135]

134 Gray, 'Clausewitz Rules', pp. 181–2.
135 *On War*, VIII, 2, p. 702.

Chapter 4

Politics: War as an Instrument

This chapter is the first of the three core chapters of the book which explore, in turn, the individual elements of the primary trinity: politics, chance and passion. Each of these chapters follows a similar structure, comprised of three main sections: Influences, Explorations and Implications.

Many commentators suggest that Clausewitz thought war was entirely governed by reason and controlled by the dictates of policy or 'interest'. According to Creveld, Clausewitz viewed war as little more than a 'rational instrument for the attainment of rational social ends'.[1] Booth has described a Clausewitzian 'political philosophy of war' which is purely rational and instrumental.[2] Even proponents of Clausewitz are guilty of overplaying this aspect of his thought through attempts to transform his insights into 'normative doctrine'[3] and overemphasising the role rational policy plays in his theory. Such interpretations contain seeds of truth, but only to a point. The idea of rationality is clearly expressed in the trinity when Clausewitz states that war is composed of 'an element of subordination, as an instrument of policy, which makes it subject to reason alone'.[4] Regardless of the somewhat misleading translation of this sentence,[5] how could an author who had witnessed first-hand the horrors and brutality of war, such as Napoleon's dreadful retreat across the Berezina, and who had seen his government adopt and cling to such misconceived policies, come to such a seemingly cold conclusion?

Undoubtedly Clausewitz thought, in both a moral and didactic sense, that the use of force *ought* to be subordinated to policy. Yet, these assertions greatly distort Clausewitz's contribution. Clausewitz's ideas are more nuanced than any crude depictions of strict political rationalism would suggest. Even a cursory acquaintance with his work would reveal the prominent role that emotional and contingent factors play in his theory. Indeed, Clausewitz would no doubt

1 Martin van Creveld, *The Transformation of War* (New York: Free Press, 1991), p. 179; and, *The Culture of War* (New York: Ballantine Books, 2008).

2 See Booth, *Strategy and Ethnocentrism* (London: Croom Helm, 1979), p. 74.

3 Antulio Echevarria, *Clausewitz and Contemporary War* (Oxford: Oxford University Press, 2007), p. 85.

4 Carl von Clausewitz, *On War*, trans. and ed. Michael Howard and Peter Paret (New York: Everyman's Library, 1993), I, 1, § 28, p. 101.

5 Bassford suggests it should read: 'pure reason'. 'Reason alone' is inaccurate because there are clearly two other dimensions in the trinity. Christopher Bassford, 'The Primacy of Policy and the 'Trinity' in Clausewitz's Mature Thought', in Hew Strachan and Andreas Herberg-Rothe (eds), *Clausewitz in the Twenty-First Century* (Oxford: Oxford University Press, 2007), p. 77.

empathise with Spinoza's view that those who think men 'can ever be induced to live according to the bare dictate of reason, must be dreaming of the poetic golden age, or of a stage-play'.[6] Indeed, Clausewitz believed that logic often came to a halt in the great labyrinth of war.[7] There is no simple explanation of the manner in which the political element fits into his theory; no formulaic or linear characterisation will suffice.

Part of the reason for misinterpretations relating to this tendency is the complexity of the subject itself, combined with its somewhat limited and disjointed presentation in *On War*. This may be surprising to some given the clear importance he attached to the idea, the emphasis placed on it by his critics and that it now essentially defines Clausewitz's modern relevance. These deficiencies may in part be due to his untimely death, but we must also recognise his work for what it is. As this chapter will show, ideas of politics, policy and reason hold a number of differing implications in terms of their relationship to one another and their influence on war. Moreover, the terms under discussion are themselves complex and contested. We must seek to understand better the ways in which these ideas developed in Clausewitz's thought and the way in which he believed they operated in relation to war. In this chapter we explore some of the major influences on his thought relating to the subject. This will provide the foundation for a more detailed analysis in the second section, before moving on to consider some of the theoretical implications deriving from the analysis.

Influences

Determining the specific influences that contributed to Clausewitz's insights regarding this tendency is a difficult yet not entirely futile task. The purpose here is not to propose any direct correlation between influence and idea. Rather, we will explore, in a somewhat impressionistic manner, a number of areas and suggest ways in which Clausewitz was guided in his thinking by the broad intellectual climate, existing ideas and his observations on the events of the day.

Clausewitz's precursors

Given that today the name of Clausewitz is almost synonymous with the idea of war as an instrument of policy, it is easy to overlook the fact that a number of theorists prior to and contemporary with him had contemplated war's connection to politics. The notion that war was somehow related to the realm of politics was not an original idea. Smith has noted that, 'the idea of war as an instrument of policy is most closely associated with Clausewitz but was already familiar in the

6 Spinoza quoted in Kenneth N. Waltz, *Man, the State, and War: a Theoretical Analysis* (New York: Columbia University Press, 2001), p. 24.

7 *On War*, VIII, 2, p. 701.

eighteenth century'.[8] Clausewitz himself accepts that the idea was more or less common-knowledge: '*It is of course well-known* that the only source of war is politics – the intercourse of governments and peoples'.[9] Echevarria notes that the contemporary Prussian *Officer's Handbook* written by August Rühle von Lilienstern had it that '"war as a whole always has an ultimate political purpose", it is, in fact, undertaken "to realise the political purpose on which the state decided in view of the nation's internal and external conditions"'.[10]

Enlightenment military theorists had focused almost exclusively on strictly military matters, considering war a subject area worthy of study as a separate discipline from that of politics – as a discrete and bounded system – yet this is not to suggest they were unaware of connections between the two. Moreover, some theorists had begun to grasp at the ideas that Clausewitz would come to present in a more thorough manner. In this respect, there was a relatively sturdy foundation of thought to draw on: his ideas did not emerge in an intellectual vacuum. The fact that Clausewitz only truly determined the precise role of politics in his final years is reason to wonder whether he could have reached those conclusions before his untimely death without such a foundation of existing ideas. The point, however, is to emphasise that he was not working from a *tabula rasa*, even if his ideas would go deeper than existing conceptions.

A theme which emerges from earlier theories is that war is treated as something that arises out of political situations, yet which – to adopt Clausewitz's metaphor of politics as the 'womb of war'[11] – orphans its child and leaves it to its own fate. This perspective allowed thinkers to conclude that mastery of war was vital if political interests were to be served, but they did not explore how one interacted with the other. Montecuccoli, for instance, thought it vital to develop the science of war because it was of the utmost political importance.[12] The first book of his *Treatise on War* was devoted to an extensive study of the nature and political context of war. For him, the purpose of war was a favourable peace – an idea common since Aristotle claimed that we 'make war in order that we may live at peace'.[13] Similarly, Friedrich von Lossau, a reformer and member of the *Militarische Gesellschaft*, of which Clausewitz was a member, held that, 'Wars are therefore the exterior means of states to achieve by violence what they cannot achieve by peaceful means'.[14]

Of Clausewitz's contemporaries, Jomini was perhaps the most advanced in his understanding on this subject. At the beginning of *The Art of War* we are presented

8 Hugh Smith, *On Clausewitz* (Hampshire: Palgrave Macmillan, 2005), p. 100.

9 *On War*, VIII, 6B, p. 731. Emphasis added.

10 Quoted in Echevarria, *Clausewitz*, p. 90.

11 *On War*, II, 3, p. 173.

12 Azar Gat, *A History of Military Thought: From the Enlightenment to the Cold War* (Oxford: Oxford University Press, 2001), p. 22.

13 Aristotle, *Ethics*, trans. J. A. K. Thomson (London: Penguin, 1976), p. 329.

14 Quoted in Smith, *On Clausewitz*, p. 100.

with a typology of war as shaped by distinctive political conditions and motives.[15] Yet, the extent of Jomini's observations on the impact of policy was that 'these different kinds of war influence in some degree the nature and extent of the efforts and operations necessary for the proposed end'.[16] Jomini's discussion appears to offer a route into deeper analysis, yet this promise fades and the influence of politics is largely absent from his subsequent theory despite some partial, anecdotal references: the autonomy of the military sphere is emphasised. As Clausewitz held, knowledge of political 'type' might certainly help us understand individual wars, but is of little use to general theory, other than to show that there are indeed many different forms of political motivation that give rise to war. In Jomini, as in the work of earlier thinkers, we do not see the same level of depth and intricacy that we do in Clausewitz.

Accepting what *is* contained in these works, they can be clearly distinguished from Clausewitz in three respects: they generally treat war as an autonomous phenomenon capable of study in its own right; following from this, they do not explore the nature of the relationship and *dynamic interaction* between the two in great detail; and finally, while groping towards the instrumental nature of war they do not explore the implications of this insight in terms of broader military and strategic matters.

Machiavelli and the realist tradition

Aside from contemporaneous theorists, there was another rich vein of thought that had a much deeper lineage and which impressed Clausewitz greatly. This was the broad tradition of what we now know as classical realism, the pre-eminent figures of which were Thucydides and Machiavelli. This represented a powerful body of thought within which the use of force in political affairs was a principal concern. Although there is no direct evidence of Clausewitz's familiarity with Thucydides, his fascination with Machiavelli is undoubted. We will thus explore Clausewitz's debt to this tradition through the prism of Machiavelli's writings.

Machiavelli's belief that war is essentially a branch of politics greatly impressed Clausewitz.[17] The great Florentine, along with other thinkers in the realist mould, emphasised the idea that war was an inescapable feature of the relations between political communities in a hostile environment that resembled the imagined human state of nature. Each was compelled to fend for itself in the absence of an overarching authority and, in the interest of security and self-preservation, others must eternally be considered as potential threats.[18] Likewise, Clausewitz believed

15 Antoine Henri de Jomini, *The Art of War*, trans. G. H. Mendell and W. P. Craighill (Texas: El Paso Norte Press, 2005), pp. 10–29.

16 Jomini, *Art of War*, pp. 10–11.

17 Bernard Brodie, *War and Politics* (London: Cassell & Company Ltd, 1974), p. 438.

18 David Boucher, *Political Theories of International Relations* (Oxford: Oxford University Press, 1998), p. 30.

'that the international arena was dominated by the behaviour of sovereign states guided by considerations of *raison d'etat* in which power played the major role'.[19] This was the basic reason for Clausewitz's 'affinity with a writer who insisted that, above all, the state was an institution created and maintained by the realistic use of force'.[20] In Machiavelli's writings war is presented as an instrument; one of the occasionally vital means which states employ to attain their ends.

Machiavelli wrote at a time of great political and military turbulence – somewhat akin to that faced by Prussia in the early nineteenth century – when the established patterns of Italian politics were under threat internally and externally. For both Machiavelli and Clausewitz the priority for government was positive action to ensure the security of the state, which could only be achieved through a combination of astute political manoeuvring and maintaining military effectiveness. Although Clausewitz did not follow Machiavelli in his harsh principles for consolidating domestic authority,[21] he certainly accepted that governments must be prepared to use force to ensure their security and independence when threatened. Machiavelli, drawing on classical texts, held that military success was greatly dependent on political conditions: a democratic civil society could produce a citizen army – as opposed to a professional mercenary force – with superior morale and, hence, enhanced battlefield effectiveness. Citizens' experience of fighting for the community would, in turn, strengthen the virtuous bonds of civic unity. For Machiavelli, the good of the state and military effectiveness, politics and war, were intricately entwined in a virtuous dialectical relationship.[22] Clausewitz strongly sympathised with the idea that they could be mutually supportive in this way.[23]

The classical realist tradition certainly provided the parameters in which Clausewitz's more detailed exploration of the nature of the relationship between war and politics could coalesce. Reading the works of such thinkers allowed him to conceive of war from the perspective of politics and to contemplate its implications. Machiavelli provided Clausewitz with profound insights into, not only why war is an inescapable feature of relationships between independent groups, but also the more subtle interconnections between civil and military affairs. Yet, while war had been a central concern of realist thinkers, there was a limit to their thought on the subject. Their emphasis had been on the political conditions which make

19 Gat, *Military Thought*, p. 246.

20 Peter Paret, *Clausewitz and the State: The Man, His Theories, and His Times* (Princeton: Princeton University Press, 1985), p. 172.

21 'What he says about the prince's policies toward their subjects is certainly largely outdated'. Carl von Clausewitz, 'Notes on History and Politics', in Carl von Clausewitz, *Historical and Political Writings*, ed. and trans. Paret, Peter and Moran, Daniel (New Jersey: Princeton University Press, 1992), p. 268.

22 Neal Wood, 'Introduction' in Niccolò Machiavelli, *The Art of War*, trans. Ellis Farneworth (Cambridge, Massachusetts: Da Capo Press, 2001), pp. xlvii–liii.

23 Carl von Clausewitz, 'Letter to Fichte', in Clausewitz, *Writings*, pp. 282–3.

war an inevitability and why it should be seen as a necessary instrument. Despite Machiavelli's extensive writing on military affairs – most significantly in *The Art of War* – he treats the actual conduct of war in a largely autonomous manner and his emphasis was on the political implications of military affairs. Clausewitz's analysis approached the subject from the reverse perspective: he endeavoured to understand how politics impacted on war.

Events of the day, experiences and historical analysis

That political factors were a running concern for Clausewitz can partly be attributed to the ideas contained in some of the existing works on military theory, but additionally we cannot discount Clausewitz's own experiences and his interpretations of the events of his lifetime as important influences. As Brodie observes, in this sense he certainly 'had some bitter and extraordinary personal experiences to help him along the way'.[24] Combined with his historical analyses, Clausewitz began to see more clearly the nature of the interaction between war and politics.

From an early age, Clausewitz had seen how military effectiveness could be fatally undermined without the correct political basis. At age thirteen, he witnessed the failure of the First Coalition to follow up the successful siege of Mainz in 1793 due to 'Prusso-Austrian political arguments',[25] he observed the timid and irresolute Frederick William II pass up the opportunity of joining the Second Coalition and in 1805 Clausewitz saw the new sovereign, King Frederick William III, cling to his neutrality while Napoleon, in the campaign culminating with Austerlitz, defeated Austria and Russia. Finding himself facing Napoleon alone, the King could not accept the Emperor's new demands and foolishly declared war. If he had read the political situation correctly, timely action in unison with Austria and Russia may have averted the calamity of 1806. But the political failure ran deeper than that. The inadequacies that would be revealed on the battlefields of Jena and Auerstedt in 1806 were not purely military, but resulted from a failure to recognise and adapt to changed political conditions. Prussia did not appreciate the consequences of the Revolution in France, particularly with regard to the new vigour it instilled in its armed forces. Prussia fought with an army perhaps suited to a form of limited Frederickan war, yet entirely unsuited to facing the 'untrammelled' war practised by Napoleon. Clausewitz asks 'would a purely military view of war have enabled anyone to detect these faults and cure them? It would not'.[26]

More generally, over the course of the Napoleonic period, Clausewitz could observe how the states of Europe had individually and repeatedly failed to achieve anything like the political unity that could have delivered strategic success.

24 Brodie, *War and Politics*, p. 441.

25 Roger Parkinson, *Clausewitz: A Biography* (New York: Stein and Day, 1979), p. 27.

26 *On War*, VIII, 6B, p. 736.

Meanwhile, Napoleon had certainly provided Clausewitz with a master-class in employing war to secure political power and of reading the intentions and interests of his enemies to play them off against one another. Yet, paradoxically, it was Napoleon's ultimate strategic failure, culminating in Waterloo, which provided Clausewitz with the clearest example of the importance of politics. His stunning military victories may have allowed him to dominate vast swathes of Europe, but his ultimate failure was political.[27]

Napoleon's battlefield victories were never politically decisive enough to lay the foundations for a lasting peace. Regardless of his military genius, his policy objectives were unrealistic and he failed to understand that peace must be accepted by the militarily defeated, otherwise there always remains the potential for renewed resistance: the 'defeated state often considers the outcome merely as a transitory evil, for which some remedy may still be found in political conditions at a later date'.[28] Napoleon refused compromise settlements and he never grasped that it was insufficient to simply dictate terms, short of imposing a 'Carthaginian Peace' on his enemies. European statesmen came to realise that coexistence with Napoleon was impossible; he had to be defeated by politically united military action. So, embedded in his great military victories were the seeds of his own downfall.[29]

Clausewitz could of course contrast the Napoleonic performance with the example of Frederick the Great who was capable of extreme vigour when necessary, but was then content to revert back to 'a state of calm oscillation, always ready to adjust to the smallest shift in the political situation'.[30] This led Clausewitz to conclude that an essential element of military genius was a 'thorough grasp of national policy' whereby the commander 'is aware of the entire political situation' and also 'knows exactly how much he can achieve with the means at his disposal'.[31]

Recognition of the importance of politics is also reflected in Clausewitz's historical works. His scholarly professionalism and commitment to objectivity compelled him to always trace events back to their prominent causes; to follow the threads of action back to their original motive. He notes that 'a means may be evaluated, not merely with respect to its immediate end: that end itself should be appraised as a means for the next and higher one … In many cases, particularly those involving great and decisive actions, the analysis must extend to the *ultimate objective*, which is to bring about peace'.[32] Consequently, his numerous campaign analyses were inevitably drawn away from tactical matters toward fundamental political problems and he always kept the political elements basic to every conflict

27 Colin S. Gray, *War, Peace and International Relations: An Introduction to Strategic History* (Abingdon: Routledge, 2007), p. 7, pp. 39–40, and p. 46.

28 *On War*, I, 1, § 9, p. 89.

29 Esdaile, *Napoleon's Wars*, p. 109.

30 *On War*, III, 1, p. 210.

31 *On War*, I, 3, p. 130.

32 *On War*, II, 5, p. 184.

steadily in view.[33] Through this process of intense historical study, Clausewitz would time and again be forced to contemplate the extent to which military events, engagements won and lost or issues pertaining to military administration were decided or at least influenced by political considerations. These analyses provided Clausewitz with a rich empirical basis on which to develop his ideas in a more theoretical fashion.

Explorations

So, through the ideas of earlier theorists, his own direct experiences, historical analysis and his reading of the events of the age, Clausewitz came to understand that politics was inextricably intertwined with the use of force and that their connection could not be overlooked in theory: but how exactly were they related? His theoretical antecedents had grasped at the linkages, but failed to follow them through or adequately incorporate them into their systems.

The overriding focus of Clausewitz's studies was that activity to which his whole professional life had been devoted: war. The closest Clausewitz came to a political appointment was his ultimately unsuccessful campaign to be appointed Prussian minister to the Court of St James in London[34] and his only diplomatic experience being the part he played in the Convention of Tauroggen, which facilitated the defection of General Yorck from the French at the end of 1812. Nevertheless, Clausewitz could not help but be repeatedly drawn towards political issues and problems. From an early age, Clausewitz had, sometimes in extremely personal ways, come face to face with the interaction between politics and war and expressed his personal political views on the pressing concerns of the day in forcible terms.[35] This is not to mention the simple fact of his 'abiding fascination with political issues and questions of foreign policy'.[36] If there was ever anyone well placed to explore the crucial relationship between war and politics, it was him.[37]

Power, politics and policy

As Bassford rightly points out, 'telling students that war is an expression of X, without defining X, gets them nowhere'.[38] Politics can mean different things to different people and it is true that Clausewitz never explicitly supplies a workable

---

33 Paret, *Clausewitz*, p. 203.

34 See Parkinson, *Clausewitz*, pp. 304–5.

35 Most notably in his 'Political Declaration', in Clausewitz, *Writings*, pp. 287–303.

36 Paret, *Understanding*, p. 178.

37 Paret notes the irony that it was a soldier who would first explicitly lay down the political nature of war. Paret, *Understanding*, p. 19.

38 Bassford, 'The Primacy of Policy', p. 83.

definition or firm guidance as to when and where he is differentiating between, what in English we would term, politics and policy, but which in German is covered by the single term, *politik*. This can lead to considerable confusion in our understanding, therefore it is necessary to provide more concrete conceptions. Clausewitz perhaps used the term quite loosely and it certainly conveys different, albeit often closely related meanings in different parts of the work.

Politics, at base, is about relative power. The concept of power has become so overused as to almost render it devoid of meaning and it is too sensitive to be encapsulated by any succinct definition.[39] An excellent starting point however is provided by Howard who states that, 'Power, in itself, is something morally neutral, being no more than the capacity of individuals and groups to control and organise their environment to conform with their physical requirements or their code of moral values'.[40] Politics is best conceived as a constant relational *process* whereby power is distributed within and between social groups. This process determines relative levels of influence that different actors have over others: it is the capacity to get others to do what you want and prevent them from doing what you do not want. As Bassford notes, it is in the nature of politics for it to be in constant and dynamic flux, 'always involving give and take, interaction, competition, and struggle'.[41] This process can take many forms: it can be either predominantly material or psychological, coercive or persuasive, conflictual or consensual or any mixture of these. For instance, power can be secured through brute force, cultural values, religious edict, traditional authority or established processes.

Politics itself, understood as a ceaseless, multilateral and changeable process of the distribution of relative power, is not a rational process in the sense of necessarily being geared towards some overall end, although it is comprised of individual actors seeking to achieve ends rationally. In terms of its application, power can be separated into those actions which seek to *compel* another to do something they otherwise would not have done or to *deter* them from doing what they would liked to have done.[42] Various tools can be used to achieve these results, such as a combination of so-called 'sticks and carrots' and these can take more active or passive forms. So, power can be secured, generated or expressed in countless ways, be it economic, diplomatic, legal, demographic, ideational, cultural or, indeed, military. All these elements of power are inherently intertwined and any realistic analysis must consider them simultaneously.

While some elements of power are material and obvious, such as military hardware, others are deeply psychological and intangible. Estimations not only of

39 Chris Brown, *Understanding International Relations* (Hampshire: Palgrave, 2001), pp. 88–9.

40 Michael Howard, 'Military Power and International Relations', *International Affairs*, Vol. 85, No. 1, January 2009, p. 149.

41 Bassford, 'Primacy of Policy', p. 85.

42 Brown, *Understanding*, pp. 90–91.

others' power, but of one's own is fraught with difficulties. War is an activity which distributes power primarily through physical coercion, although effective military power, as Clausewitz well knew, is often reliant on non-material factors and complemented by other forms of power simultaneously. The quality of command, morale and doctrine greatly contribute to measures of actual capability.[43] Power can also be greatly bolstered by other factors such as political alliances, personality,[44] and credibility based on past behaviour, or even asymmetric sources of strength such as diplomatic and military surprise or the adoption of a defensive posture. Politics is thus a necessarily multifaceted, changeable and complex phenomenon: a view strongly apparent in Clausewitz's work.

Even if one could freeze time and examine a particular group in isolation, no conclusive quantification of power would be possible. Moreover, it is precisely the relational nature of power that is most important – the attributes mentioned above have little meaning other than in relation to those against which it is directed.[45] These caveats do not render power any less real, they simply reveal its mutable and intangible features. This is particularly the case for entities in conditions of anarchy – such as for states in the international system – where, compared to an ordered domestic political context, the stabilising effect of law, legitimacy and morality is less apparent. In such situations, or when domestic political structures disintegrate, the use of force often becomes an ever-present possibility, if not a permanent reality and as Clausewitz recognised, 'certain self-imposed, imperceptible limitations hardly worth mentioning, known as international law and custom' scarcely ameliorate this condition.[46]

Policy can be viewed as a particular actor's subjective external orientation within this wider political realm. The two overlap where actions based on a particular policy become part of a political process by way of their interaction with other actors. Thus, while policy only truly has meaning in relation to wider circumstances, it also describes something particular by virtue of its being 'unilateral and rational' and to the extent that it represents 'a conscious effort by one entity in the political arena to bend its own power to the accomplishment of some purpose – some positive object, perhaps, or merely the continuation of its own power'.[47] Policy can of course, be 'at odds with politics'[48] to the extent that it often adopts courses of action ill-suited to prevailing political circumstances. Policy does not necessarily read political situations correctly, but nevertheless, it is

43 See Stephen Biddle in his *Military Power: Explaining Victory and Defeat in Modern Battle* (Princeton: Princeton University Press, 2004).

44 Wellington remarked that Napoleon was worth 40,000 men in the field. Peter Browning, *The Changing Nature of Warfare: The Development of Land Warfare from 1792 to 1945* (Cambridge: Cambridge University Press, 2002), p. 32.

45 Brown, *Understanding*, p. 92.

46 *On War*, I, 1, § 2, p. 83.

47 Bassford, 'Primacy of Policy', p. 85.

48 Echevarria, *Clausewitz*, p. 85.

an inextricable part of the wider political process and, in implementation, it creates new political realities.

While policy is unilateral we must be mindful not to view it simply as a rigid 'given', a static statement of intention or an immovable reality such as a written manifesto. Behind the concept of policy are real people, reacting to events, considering sacrifices made and risks to be weighed. Importantly, policy implies action: it is the will that sets the group on a particular course, it determines movement along new bearings and has the potential for adaptability and change. Although we may be able to identify a pre-eminent overarching objective, policy itself can be plural – it can be comprised of multiple, simultaneous or subsidiary ends, some of which may prove incompatible with others; just as individuals can be driven by numerous and conflicting motives. So, policy relates to the ends of a community but, it must also be seen as a more dynamic, multifaceted phenomenon, actively shaping parameters of action, determining resources to be employed and generally acting as a conscious force which orients and drives political collectivities in various directions.

Problems with policy in Clausewitz's work

In the light of this brief discussion, some immediate problems in relation to Clausewitz's ideas arise. Three issues in particular require consideration: policy is presented in a unitary fashion; the term policy is narrowly associated with the modern state; and there is an underlying assumption of policy rationality and reasonableness. We will examine each in turn.

First, it is true that Clausewitz treats policy in a distinctly unitary fashion, largely skating over political processes internal to its formulation. He states that the 'aim of policy is to unify and reconcile all aspects of internal administration as well as of spiritual values … it is simply the trustee for all these interests against other states'.[49] This reflects his paternalistic view of politics, common at the time, which saw the state leadership as the natural repository for political decision.[50] He even describes policy as 'the intelligence of the personified State'.[51] Clausewitz thus downplays the hard-fought political battles that exist at all stages of policy formulation.[52] Yet, during his career, Clausewitz himself 'bitterly opposed the political aims and even the declared policy of the king'[53] and must have been aware that political contention meant decisions were by no means predetermined. The dynamics of internal politics may be crucial as they can powerfully shape the ends, choices and behaviour of belligerents.

49 *On War*, VIII, 6B, p. 733.

50 Gat, *Military Thought*, p. 247.

51 Carl von Clausewitz, *On War*, trans. J. J. Graham and ed. F. N. Maude (London: Penguin, 1982), p. 120.

52 See Echevarria, *Clausewitz*, p. 93.

53 Gat, *Military Thought*, p. 250.

The precise locus of decision-making in relation to the use of force can be difficult to determine, yet basic distinctions help us recognise what we might call 'the realm of war': political battles certainly take place within groups, but war is conducted between at least partially unified entities whose policy involves the threat or use of force directed against others. In this sense, Clausewitz's assumptions about *a single dominant policy* may not be as misleading as some have held; we must just be aware that it shrouds underlying divisions and conflicts of interest that continually shape policy directions. In many cases, internal divisions develop into outright conflict, which greatly alter the complexion of war as a whole. Also, this conception does not preclude recognition of the fact that some actors within groups betray their own side, collaborate and conspire with the enemy or, in some cases, defect entirely. War, up close, is replete with such behaviour, but as we zoom out, the basic divisions and confrontations in the realm of war are usually clear.

The second criticism relates to the very use of the term policy. Some commentators adopt parochial conceptions of policy as being solely the preserve of modern states and, by extension, the mistaken conclusion that Clausewitz's dictum 'means nothing more or less than that [war] represents an instrument in the hands of the state'.[54] Yet, policy need not be understood in this historically confined sense and it 'goes without saying that political energy is created not only by the state, and military power is often associated with other political interests or factions'.[55] Of course, Clausewitz had the modern state at the forefront of his mind, yet he was well aware that other war-making peoples have existed – from empire to roving horde[56] – that possessed nothing like the level of institutional sophistication of modern states, yet nevertheless identified objectives and conducted war in pursuit of them. Alternative terminology is arguably limited. 'Purpose', for instance, is vague and can encompass many things not related to issues of power or politics. Phrases such as 'political purpose' or 'political objective' are acceptable, but policy – understood as 'a planned line of conduct in the light of which individual decisions are made and co-ordination achieved'[57] – remains a useful shorthand.

Third, war is said to be subject to reason *because* it is an instrument of policy: the implication being that policy is somehow a natural repository of reason. What is Clausewitz suggesting here? At base, reason is a mode of thought peculiar to humans due to their ability to speculate and form logical judgements. In practical contexts, reason is associated with rationality; that is, behaviour that seeks to tailor means to ends.[58] We therefore speak of rationality when humans coherently

54 Creveld, *Transformation*, p. 125.

55 Paret, *Understanding*, p. 10.

56 Echevarria, *Clausewitz*, p. 84.

57 Mike Smith, 'Strategy in an Age of "Low-Intensity" Warfare: Why Clausewitz is Still More Relevant than his Critics', in Duyvesteyn and Angstrom, *Rethinking*, p. 35.

58 Michael Oakeshott, *Rationalism in Politics and Other Essays* (Indianapolis: Liberty Fund, 1991), p. 99.

identify objectives, define aims to achieve, determine methods to follow and develop criteria to judge achievement of objectives. Importantly, neither eventual success nor the precise nature of the objectives are vital conditions: rather it is the form of the behaviour, not its outcome or character that defines its rationality.[59] So, an essential precondition for rational action is purpose. War will contain an element of reason in so far as it is a social phenomenon resulting from conscious behaviour directed toward some aim. That humans sometimes fail to achieve their ends, either because they were objectively unattainable or because they apply inappropriate means, does not necessarily negate the existence of rationality.

Clausewitz's aim was not to measure behaviour in war against some perfect standard of reason and he is certainly not suggesting that all groups always act perfectly rationally. He was well aware of the devastating follies which leaders commit, which are often, as Howard argues, due to a 'superabundance of analytical rationality'.[60] In accepting an association of reason with policy, Clausewitz is not claiming that all policy is necessarily correct or sensible: we should not confuse rationality with wisdom in policy.[61] Nor was he claiming that policy would necessarily be *reasonable*, at least as far as the purpose of any war can be judged against universal standards of morality or the laws and customs of the 'war convention'.[62] As Clausewitz knew well, policy 'can err, subserve the ambitions, private interests, and vanity of those in power'.[63]

Creveld describes the idea that war must always be for some purpose as the 'original sin' of modern strategic thinking, due primarily to Clausewitz's influence. War for Creveld cannot be rational because 'it is absurd for a person to die for somebody or something else' and 'to die for one's own interest is almost equally absurd'. He argues people must fight 'only to the extent that they experience war itself and everything pertaining to it as an end'.[64] This is misleading in a number of respects. First, his argument rests on the contention that individuals entering into war expect that they *will* die. Yet this is not necessarily the case, so surely it can be rational to *risk* death for certain objectives given the possible consequences of inaction. Second, Creveld underestimates individuals' conception of themselves as part of a community; that humans fight to defend their kin-group because they are subliminally aware of their own weakness without its protection. This has deep evolutionary origins, and explains how group survival can become synonymous

59 Oakeshott, *Rationalism*, p. 100.

60 Michael Howard, *The Causes of War* (Cambridge, Massachusetts: Harvard University Press, 1983), p. 14.

61 Bassford, 'Primacy of Policy' p. 85.

62 For an interesting discussion on this point see, Mary Kaldor, 'Inconclusive Wars: Is Clausewitz Still Relevant in these Global Times', *Global Policy*, Vol. 1, Issue 3, October 2010, pp. 278–9.

63 *On War*, VIII, 6B, p. 733.

64 Creveld, *Transformation*, pp. 190–91.

with individual survival.[65] As Aristotle held, the individual is crucially dependent on the community: 'man is by nature a political animal'.[66] Finally, Creveld's is a predominantly individualistic argument for an activity that is collective by definition. The reasons why any particular individual is fighting are likely to differ from those of the social group. That war becomes an end in itself for some of those caught up in it, even the leadership, does not seriously jeopardise Clausewitz's conception of war as collective instrumental and purposeful behaviour. We must now consider this latter important point, as formulated by Clausewitz, in detail.

War's subordination as an instrument of policy

That war is an instrument of policy has become something of a truism, almost to the point of cliché, in Western strategic literature, regardless of how well the implications and complexities of the idea are understood. Its ubiquity can in large part be attributed to Clausewitz: often direct reference is made to *On War* whenever this principle is outlined.[67] The idea is commonly used out of the context which Clausewitz intended, thus draining it of much of its meaning. Also, it is often mistakenly portrayed as representing the totality of his theorising. Its most common modern usage is as a prescriptive device; one especially suited to liberal democracies where the subordination of the military to civilian control is deemed a vital component of a properly constituted modern state, particularly in the nuclear age. The complexities of the concept are often diluted in the interest of doctrinal precision and pedagogical clarity. Given the profusion of critiques in relation to this aspect of Clausewitz's thought, this situation will not suffice.

However, it is not sufficient to defend this aspect of Clausewitz's thought through insistence: by *simply repeating it* in the face of counterarguments. If anything, such insistence casts greater doubt on its veracity. We must seek to reveal why it is mistaken to assume that this conception was simply 'invented', but is indeed a universal element of war. A more robust explanation is required: what is actually meant by the notion that war is an instrument of policy?

At a basic level, Clausewitz is making the point that the use of military force is a means to a higher end and the end in question is the political object: war is a tool which policy utilises to achieve its objectives and, as such, has a measure of rational utility. To make this claim may appear 'ruthless, cold-blooded, militaristic and abhorrent'.[68] Yet, from the outset we should be clear that his assertion is

65 See Ehrenreich, *Blood Rites*.

66 Aristotle, *Politics*, pp. 59–61. He posits that to be completely independent of society is to be either subhuman or superhuman, a dumb animal or God.

67 Howard notes that the 'popular view which links his name with the phrase is not at fault. This was no casual *obiter dictum*, but the essence of his thought'. Howard, 'War as an instrument of policy', in Herbert Butterfield and Martin Wight (eds), *Diplomatic Investigations* (London: George Allen & Unwin, 1966), p. 193.

68 Creveld, *Transformation*, p. 126.

descriptive, not prescriptive.[69] This is confused by the fact that Clausewitz often draws prescriptive conclusions from the observation and the two perspectives are often presented close together in the text. It is ironic that, what can at first sight appear to be a morally repugnant statement – because it suggests Clausewitz viewed the resort to force as an 'entirely routine extension of unilateral state policy' as if the 'policy of a state is incomplete unless war is one of its instruments'[70] – actually led him to conclude that war *ought*, in a practical and moral sense, to be subject to policy otherwise it becomes 'something pointless and devoid of sense'.[71]

So, what are the implications of insisting that war is subordinate to policy? It is useful to explore the extreme boundaries of what this conception, taken alone, entails. In this light, just as any instrument is shaped by the purpose for which it exists, so too will war be shaped by its: war 'is controlled by its political object',[72] which 'will set its course, prescribe the scale of means and effort which is required, and makes its influence felt throughout down to the smallest operational detail'.[73] The purpose for which the use of force is intended will be the major determinant of the course and character of the war. This idea is clearly reflected in Clausewitz's work where he discusses the relationship between purpose, aim and means. He describes distinctly rational chains of action and feedback, which establish a purpose to be achieved, a military aim that will best serve the purpose and the selection of means most appropriate to attainment of the military aim.

Through this hierarchically structured logic, we can expect the means used in war to remain firmly under the control of policy. We should be able to explain, in broad terms, the actions of the individual soldier by the overarching demands of policy, because all parts stand in logical relation to it. The control of policy might manifest itself in, for instance, setting geographical limits to an army's movements or establishing the appropriate moment to seek a negotiated settlement. This perspective does not necessarily downplay military considerations; war 'is entitled to require that the trend and designs of policy shall not be inconsistent with these means'. Policy must understand the capabilities, limits, strengths and weaknesses of the instrument it employs, yet the crucial point is that all military action is ultimately geared towards political objectives: military concerns 'will never do more than modify them'.[74]

In this pure conception, all action rationally relates to the given purpose. As Clausewitz notes, because 'it is policy that has created war' it is only natural that it remains subordinate to the 'guiding intelligence that brought it into existence.[75] In one of his most rationalistic passages, he notes that 'the value of the object must

69 Howard, 'War as an Instrument of Policy', p. 193.
70 Howard, 'War as an Instrument of Policy', p. 194.
71 *On War*, VIII, 6B, p. 732.
72 *On War*, I, 2, p. 104.
73 *On War*, VIII, 2, p. 700.
74 *On War*, I, 1, § 24, p. 99.
75 *On War*, VIII, 6B, p. 733.

determine the sacrifices to be made for it in *magnitude* and also in *duration*. Once the expenditure of effort exceeds the value of the political object, the object must be renounced and peace must follow'.[76] The assumption is that those leading the war have chosen the most appropriate means to attain the desired end and that the latter will remain the principal criterion of efforts to be made and resources employed. The particular militarily solution will depend on circumstances and the object in question. As Clausewitz notes, there are many potential roads to victory to choose from and a 'prince or general can best demonstrate his genius by managing a campaign exactly to suit his objectives and his resources, doing neither too much or too little'.[77]

The description above, while capturing the essence of the issue, by no means represents the extent of Clausewitz's ideas. If it did, no doubt accusations of pure rationalism would be justified. The idea of a rational process does feed into the concept of war as subordinate to policy, *but only to an extent*. Where Clausewitz discusses purpose and means, or war plans, his intention is clearly prescriptive and represents an idealised strategic construct whereby all parts gel seamlessly into a rationally directed whole. This perspective is most clear in Book 8 where he states that: 'War plans cover every aspect of war, and weave them all into a single operation that must have a single, ultimate objective in which all particular aims are reconciled'.[78]

Undoubtedly, this idea is practically seductive and reflected the state servant in Clausewitz. While his ultimate intention was to describe the objective nature of war, this did not prevent him from passionately promoting his ideas regarding how war *should* be fought. These perspectives sit together uneasily and we must be careful to distinguish between the two. Nevertheless, Clausewitz stressed the great difficulties that he knew, in reality, intervened in such theoretically neat relationships:

> The degree of force that must be used against the enemy depends on the scale of political demands on either side. These demands, so far as they are known, would show what efforts each must make; but they seldom are fully known … Nor are the situation and conditions of the belligerents alike … Just as disparate are the governments' strength of will, their character and abilities … These three considerations introduce uncertainties that make it difficult to gauge the amount of resistance to be faced and, in consequence, the means required and the objectives to be set.[79]

This reality, Clausewitz persists, means any strictly logical answer to the problem is unrealisable; indeed such reasoning may prove to be 'a most unsuitable and

76 *On War*, I, 2, p. 104.
77 *On War*, III, 1, p. 208. He probably had Frederick the Great in mind here.
78 *On War*, VIII, 2, p. 700.
79 *On War*, VIII, 3B, p. 707. Emphasis added.

awkward intellectual tool'.[80] Rather, what is required is intuitive *judgement* 'to detect the most important and decisive elements in the vast array of facts and situations'.[81] This requires decisions and behaviour not strictly derived from rational processes or abstract principles. Clausewitz certainly did not hold that any superlative standard of rational behaviour, which would always maximise the interests of policymakers through the use of force, was humanly achievable.

The form of instrumental logic described above will, Clausewitz held, be apparent in any war, but the extent of its control will be extenuated by many factors, some associated with the nature of instrumentality itself. Instruments, of course, can be used for purposes other than those that they were ostensibly intended, may be ill-suited to the tasks asked of them or be employed by the user ineffectively.

Also, as Clausewitz emphasises, the subordination of war to policy by no means entails complete control. Subordination, he explains, 'does not imply that the political aim is a tyrant'.[82] Just as a subordinate soldier can disobey commands, so too can war in relation to policy. Policy, he states, 'will permeate all military operations, and, *in so far as their violent nature will admit*, it will have a continuous influence on them'.[83] The italicised text above underlines that, as Echevarria notes, 'contrary to what many scholars have maintained, Clausewitz placed some significant limits on policy's control over military operations'.[84] Clausewitz could not be clearer in his assertion that, even though the political aim will remain the 'first consideration' because the 'prime cause of its existence will remain the supreme consideration in conducting it', war – more than any other instrument – can often escape the control of its user. However, Clausewitz argues that war usually lasts long enough for policy to maintain some control; for it to 'remain subject to the action of a superior intelligence'.[85]

Recent critiques have denied war's instrumentality and point to the cultural determinants of behaviour.[86] Such studies are misleading. It is undoubted that war develops its own peculiar culture and that this can be detrimental toward achieving stated ends, but overemphasis of this point can distract us from essentials. All rational behaviour takes place in culturally conditioned environments, for no actor operates outside of culture. To posit that rational action must be action unaffected by culture is to deny the potential for rationality. Moreover, many cultural forms can be explained in instrumental terms. Indeed, taking the longer view, dominant strategic cultures are largely reflections of success in war, the consequence of adaptation to challenges.[87]

---

80 *On War*, VIII, 2, p. 702

81 *On War*, VIII, 3B, p. 707.

82 *On War*, I, 1, § 23, p. 98.

83 *On War*, p. 99. Emphasis added.

84 Echevarria, *Clausewitz*, I, 1, § 8, p. 88.

85 *On War*, I, 1, § 23, p. 98.

86 Martin van Creveld, *The Culture of War* (New York: Ballantine Books, 2008).

87 Gray, *Modern Strategy*, pp. 146–7.

But such critiques are mistaken at a more basic level. Clausewitz certainly frequently and forcefully exposes the limits of rationality given the play of chance and irrationality, as well as the way the unique character of different societies affects decisions in war. In describing war's instrumentality, he was not suggesting that all activity in war is perfectly geared towards objectives. We must be careful not to confuse appearance with substance. Clausewitz *appears* to treat war in purely rational terms, when in fact, on closer reading, the *substance* incorporates complexity, contradiction and nuance. Prescription is confused with description, instrumentality with the limits to practical rationality, and subordination with complete control.

Nevertheless, the concept of war's subordination to policy causes us to focus on an important aspect of war relating to the extent to which, from the perspective of individual groups involved, there is a constant, albeit messy, interaction between ends and means at all levels. This weaves a loose thread of rationality through the whole tapestry, even if it may appear to be absent in some situations.[88] However, ceasing the analysis here would ignore a central idea Clausewitz wanted to convey. Indeed, mistaken interpretations often stem from a failure to explore beyond policy into the more complex realm of *politics*.

From 'subordination' to 'continuation'

As Clausewitz stressed, war is a multilateral and interactive phenomenon, so how can it be described simply as the instrument of one (or more) actor(s)? To answer this we need to return to the text and reveal the way in which policy should always be understood as one aspect of a multilateral, interactive and complex whole.[89]

At the beginning of Book 8, Chapter 6B, Clausewitz provides one of his clearest expressions of the relationship between war and politics. Over the first five paragraphs the context of his discussion is clearly about the *total political situation*, as distinct from unilateral policy. He explains how the 'source of war is politics – the intercourse of governments and peoples', that it is a 'continuation of political intercourse, with the addition of other means' and that all the salient factors that make up war are political or 'so closely connected with political activity that it is impossible to separate the two'.[90] The rest of the chapter reverts to the unilateral, instrumental perspective, which is natural given that Clausewitz is dealing with war plans. He then variously refers to 'policy', 'political decisions' and 'political objectives'.

88 For instance, it appears Japan and Germany abandoned almost all semblance of rationality in their decision to continue fighting in the closing stages of World War II.

89 The following analysis holds that the Howard/Paret translation of *On War* is accurate in its usage of the terms 'policy' and 'politics', where in German only the single term *politik* exists.

90 *On War*, VIII, 6B, pp. 731–2.

So, Clausewitz is keen to draw attention to the importance of the broad political situation out of which war arises, and the idea of instrumentality emerges so seamlessly out of the idea of interactive politics that the transition from one to the other is almost missed. As natural as that may have seemed to Clausewitz, it has caused problems for students of his ideas. This seamlessness, however, is an initial clue to his belief in the indissoluble connection between the two; that the one is implicated in the other. This is clarified by turning to the first Book of *On War*. Here policy and politics are even more intertwined and everywhere treated together. Crucially, in Section 23, Chapter 1, Clausewitz states that, 'When whole communities go to war … the reason always lies in some political situation, and the occasion is always due to some political object. War, therefore, is an act of policy'. Here, policy is presented as an intrinsic element of the political situation, essentially arising out of it. The discussion continues largely from the perspective of unilateral policy. The instrumental nature of war is reaffirmed in Section 24, but here the connection with politics is stressed: 'We see therefore that war is not merely an act of policy but a true political instrument, a continuation of *political intercourse*, carried on with other means'.[91]

War as an instrument of policy is almost nowhere presented as independent from the idea of war as a continuation of political intercourse. From an objective viewpoint, as a whole, war emerges from a political situation between belligerents and does not suspend that intercourse but carries it on with different means. To switch abruptly to the policy perspective is natural because the political situation is comprised of the interaction of the individual policies of belligerents. War is both a *continuation* of the political situation and the *instrument* employed by the groups that comprise that situation.

Once we acknowledge this, we move further from the ideal of controlled instrumentality. Indeed, the idea takes on a very different meaning in so far as it is war's very subordination to policy, and by extension politics, that makes its rational control extremely difficult. Politics as a process is not rational in itself, thus, if war is a continuation of that process, neither is it, as a whole, a rational phenomenon. Although we maintain that belligerents in war generally act according to their objectives, this occurs in a complex, multilateral and interactive environment pervaded by uncontrollable external political dynamics and chance occurrences. War is shaped not purely by the reasoned objectives and decisions of each actor, but also by the complex nature of the political dynamics that their interaction produces, which will be inherently unpredictable and non-rational.

The political logic of war

To state that war is a continuation of politics is to make a point beyond that of war's subordination or instrumentality. Continuation powerfully conveys the idea that war is itself *a form of political behaviour* and thus, even if subjective policy

91 *On War*, I, 1, § 23 and § 24, pp. 98–9. Emphasis added.

loses control of its instrument, the lines that run through war remain fundamentally political in nature. This idea cannot be overstated because, as Clausewitz states, the concept of subordination may encourage the belief that war is something entirely different to politics because,

> it is apt to be assumed that war suspends that intercourse and replaces it by a wholly different condition, ruled by no law but its own … [But] war in itself does not suspend political intercourse or change it into something entirely different. In essentials that intercourse continues, irrespective of the means it employs. The main lines along which military events progress, and to which they are restricted, are political lines that continue throughout the war into the subsequent peace.[92]

To clarify his position, Clausewitz provides a powerful metaphor: while war might have its own unique grammar, the underlying logic of war is always political. Naturally, armed groups decide how best to achieve the defeat of the enemy based on military operational realities somewhat in isolation from political concerns, and war's grammar alone, as Clausewitz conveyed through his abstract concept of 'absolute war', tends to gravitate towards the annihilation of the enemy.[93] The idea of subordination does not suggest war is entirely governed by political considerations; commanders should be free to use their judgement to determine how best to achieve the military aim demanded by policy.

So, war is a means and it has its own peculiar character – the dynamics of combat – that distinguishes it from typical political behaviour, but, as the concept of continuation conveys, at heart war *is* political behaviour. War substitutes the use of force for 'speech or writing' and represents a different expression of the thoughts of political leaders. Continuation underlines why war should never be viewed as representing a unity in itself. The idea of subordination enables us to conceive of a strong relationship, yet can be misleading as it implies the substitution of political behaviour by military force. The idea of continuation goes further, emphasising that war's complete autonomy is impossible because of their 'indissoluble connection': war is a form of political behaviour and embodies its basic logic. Both are social forces which determine the distribution of power among groups, even if the means of war – its grammar – is distinctive and possessed of its own character, dynamics and principles. The instrument of war may take many forms and serve many purposes, but its universal essence can be grasped when we understand that policy is inherently embedded in the perpetual realm of the political, within which the use of force is but a continuation of that interaction with altered means. It is in this respect that Clausewitz could state that 'all wars are things of the same nature'.[94]

---

92 *On War*, VIII, 6B, p. 731.

93 *On War*, VIII, 2, pp. 700–703.

94 *On War*, VIII, 6B, pp. 731–7.

Even if this idea is understood, there remains considerable room for confusion because history reveals countless occasions when it has been rejected by practitioners and theorists alike. In fact, the default position has arguably been one that clings to a clear separation between the two: the belief that once war begins, politics ends and success can be achieved by military means alone. This conception is mistaken, for it confuses a necessary relationship with a mere point of view. For instance, by extolling battle, Napoleon was able to secure repeated military victories, while simultaneously generating a political situation fundamentally inimical to the survival of his Empire: war will have political consequences however much practitioners may believe or desire otherwise. War may be an instrument, but if it is to prove an effective one it should be deftly handled – often alongside other tools of policy – otherwise politics has a habit of delivering its own verdict on events. As Clausewitz noted, 'War does not contain *in itself* the elements for a complete decision and final settlement'.[95]

Any particular commander may fervently subscribe to the notion that war should be fought regardless of political imperatives. This position is understandable, yet objectively irrelevant and simply reveals the way in which war can be subjectively mishandled; but the effects of such a mishandling will always ultimately be political. Outright military victory may be the most appropriate way of achieving one's object, but history shows that, equally, and frequently, it may not. Modern counterinsurgency operations reveal that the use of overwhelming force may prove enormously strategically counterproductive. As the adage rightly has it: war is too important to be left to the generals. Yet, if war is never autonomous, how exactly does politics interact with and impact on the phenomenon?

The two types of policy

So, Clausewitz believed that because war was not an autonomous phenomenon, to a considerable extent 'the nature of the political aim, the scale of the demands put forward by either side, and the total political situation of one's own side, are all factors that in practice must decisively influence the conduct of war'.[96] In what way exactly? The answer is difficult to determine and Clausewitz never completed his ideas on this subject.[97] Surveying the historical record, one is struck by the vast possibility of the forms political objectives can take. A further difficulty arises when we consider how difficult it often is to determine the true motives behind a given purpose. Indeed, Clausewitz notes that, in war, 'the facts are seldom fully

95 Clausewitz quoted in Michael I. Handel, *War, Strategy and Intelligence* (London: Routledge, 1989), p. 455. Emphasis added.

96 *On War*, VIII, 5, p. 727.

97 See Heuser, *Reading Clausewitz*, pp. 180–86; Jan Willem Honig, 'Strategy in a Post-Clausewitzian Setting' in Gert de Nooy (ed.), *The Clausewitzian Dictum and the Future of Western Military Strategy* (The Hague: Kluwer Law International, 1997), pp. 112–13.

known and the underlying motives even less so'.[98] For these reasons, there is little value in developing a schema of possible political purposes: the problem is one of explaining and generalising about political behaviour and war.[99]

Clausewitz developed a conceptual approach which allowed him to distinguish between two types of war according to the nature of their overriding purpose. In his note of 1827 he stated his intention to develop the idea that wars can essentially be either unlimited[100] – those fought 'to overthrow the enemy' and dictate terms: the 'terrible battlesword'[101] – or limited – those fought for objectives to be achieved by negotiation or settlement: 'the light, handy rapier'.[102] In reality they encompass an enormous variety 'ranging from a war of extermination down to simple armed observation'[103] and so might best be conceived as points on a spectrum. In presenting the two types, Clausewitz did not deny that 'between these two extremes lie numerous gradations'.[104] His contention was that wars fought on the basis of one or the other type would be pervaded by differing imperatives.

Wars fought for unlimited objectives, Clausewitz believed, will tend to encourage extremes. For Clausewitz, Napoleon's campaigns were examples *par excellence* of this type. Whether the rise to extremes occurs is a matter of circumstance. The enemy may rapidly succumb, but if it chooses to resist, war will lead to extremes. The sacrifices demanded of the opponent are total and so, if it wishes to prevent the loss entailed, only a correspondingly extreme effort will suffice. In such cases, war may appear more military than political, but this is only due to an elision where policy is *seemingly* effaced by the military aspect and the overthrow of the enemy essentially becomes both political purpose and military aim. Nevertheless, unlimited wars are as political as limited wars.

Instances of politically limited wars are manifold. The wars of eighteenth-century Europe are often presented as classic examples, being greatly restricted in their aims and which Clausewitz caricatured as 'a somewhat stronger form of diplomacy, a more forceful method of negotiation'.[105] Limited aims do not necessarily mean large battles will not occur or that they will be fought without ruthlessness.[106] Indeed, a modest aim may best be served by a short unforgiving

---

98 *On War*, II, 5, p. 181.

99 As Chris Brown has noted, 'classification is not the same as explanation'. Brown, *Understanding*, p. 76.

100 Unlimited war, which is real, is very different to the idea of absolute war, which is an abstraction.

101 Clausewitz, 'Note of 10 July 1827', in *On War*, p. 77.

102 *On War*, VIII, 6B, p. 732.

103 *On War*, I, 1, § 11, p. 91.

104 Howard, *Clausewitz*, p. 53.

105 *On War*, VIII, 3B, p. 713.

106 The eighteenth century was an 'age of battles' with a higher ratio of casualties to participants than in the Napoleonic Wars. Brian Bond, *The Pursuit of Victory: From Napoleon to Saddam Hussein* (Oxford: Oxford University Press, 2006), p. 14 and p. 96.

campaign to rapidly compel the enemy to negotiate.[107] Nevertheless, when engagements do occur they will tend to be restrained and there will be a greater proclivity on all sides to seek negotiations; even the victor in battle will prefer a favourable peace to risking further losses. In limited wars, 'the less will the military element's natural tendency to violence coincide with political directives … the conflict will *seem* increasingly political in character'.[108]

The central point he wished to establish was that the type of the political aim will be a powerful factor in shaping the character and conduct of war. However, understanding the objectives of either side is necessary but not sufficient. War, as Clausewitz reminds us, is akin to a duel. Thus, the political object as a 'standard of measurement' only makes sense in the 'context of the two states at war'.[109] For instance, while one side may have limited aims, there is no assurance that the enemy will follow suit and that 'the half-hearted war does not become a real war after all'.[110] The true character of any war results from the interaction of opposing purposes: it thus takes two to fight a limited war.

The idea of politically limited war enabled Clausewitz to explain why wars rarely reached the extremes suggested by logic, where extraneous factors such as friction could not fully account for limitations in reality. Whether we fight on, sue for peace through negotiations or surrender will be strongly determined by our object and the value attached to it. The interaction of these considerations on either side will be the major determinant of the continuation, scope and magnitude of hostilities.[111] Clausewitz did not, however, view policy as the sole determinant of war's character. War's own 'grammar' exerts a powerful influence. As Lawrence Freedman explains, 'Military methods must be geared not only to the political stakes but also to the capacities and methods of the adversary. The political pressures may be towards minimum force but the military pressures may point in the opposite direction'.[112]

Furthermore, the impact of the two types of policy cannot be properly conceived in isolation from the other tendencies of the trinity. So, because war is pervaded by probabilities rather than necessity, belligerents may sustain hostilities beyond what policy would rationally dictate in the hope of a favourable turn of events. Likewise, limited wars are not immune from passion. Clausewitz states that 'war and its forms result from ideas, emotions, and conditions prevailing at the time',[113] and that, 'Between two peoples and two states there can be such tensions, such

107 Bond, *Pursuit*, p. 14.

108 *On War*, I, 1, 25, p. 99. Emphasis added.

109 *On War*, I, 1, § 11, p. 90.

110 *On War*, VIII, 6A, p. 730.

111 *On War*, VIII, 6A, p. 729.

112 Lawrence Freedman, 'The Age of Liberal Wars', in David Armstrong, Theo Farrell, and Bice Maiguashca (eds), *Force and Legitimacy in World Politics* (Cambridge: Cambridge University Press, 2005), p. 103.

113 *On War*, VIII, 2, p. 702.

a mass of inflammable material, that the slightest quarrel can produce a wholly disproportionate effect – a real explosion'.[114] It is in such ways that 'the same political object can elicit differing reactions from differing peoples, and even from the same people at different times'.[115] Nevertheless, policy remains the chief arbiter of the course of war. Unlimited objectives tend to arouse and engage the emotions. Where policy aims are limited 'the emotions of the masses will be little stirred'[116] and it is likely that, in such situations, 'a policy of maximum exertion would fail because of the domestic problems it would raise'.[117]

From a strategic perspective, it is vital for those conducting war to understand the extent of the opponent's object because, if the enemy aims at your destruction, then plans for and decisions within war would entail very different considerations than if they have only limited goals. A large part of America's failure in Vietnam derived from the inability to understand that even though it may have believed it was fighting a politically limited war, its enemy certainly was not. Vietnamese forces were willing to sustain enormous casualties to secure independence, whereas mounting US casualties turned opinion against a war that was deemed unnecessary by large segments of the population.[118]

The political web of war

A major implication of the assertion that war is a continuation of politics is that the panoply of perpetually shifting relations within, between and beyond individual groups will impact on and shape the course of war. The political web within which war takes place can influence the types of policies adopted, objectives defined, decisions taken and actions initiated in war: as Echevarria notes, war and 'policy is shaped by the processes and conditions within which it is developed, in a word by politics'.[119] This fact – underrated by Clausewitz – helps account for many otherwise inexplicable events in past wars. Indeed, sometimes, action which makes sense militarily is apt to be entirely subverted by politics.

This is particularly relevant with respect to the effects arising from internal politics. Decisions taken will not always be geared towards strategic or military imperatives, but might be dictated by electoral politics, inter-service rivalry, private gain or bureaucratic considerations. In such cases, 'political considerations that override the grammar of war may run counter to the accomplishment of policy aims'.[120] A prominent instance of domestic politics driving military decisions is the phenomenon of external wars engineered by states to stave off impending domestic

114 *On War*, I, 1, § 11, p. 90.
115 *On War*, p. 90.
116 *On War*, I, 1, § p. 100.
117 *On War*, VIII, 3B, p. 707.
118 Stanley Karnow, *Vietnam: A History* (London: Pimlico, 1994).
119 Echevarria, *Clausewitz*, p. 74.
120 Echevarria, *Clausewitz*, p. 95.

insurrection.[121] Such domestic political imperatives feed into the use of military force at various levels and in too many different ways to outline comprehensively here.

Such political dynamics compound the difficulty of arriving at a clear political purpose or in ensuring military operations are conducted according to achievable objectives. Where extraneous political concerns impinge on strategic decisions, soldiers can find themselves being sacrificed for the sake of a politician's position in office or the continuance of amicable relations with allies. It is important to note, however, that apparently extraneous political concerns may serve important grand strategic purposes or ones deemed important enough to override military concerns. During the American Civil War the Union launched militarily wasteful campaigns to ensure President Lincoln's re-election, driven by the belief that ultimate victory could only be ensured with him at the helm.

Wars, or events within war, are often utilised as political tools for objectives that differ from proclaimed objectives. Whatever the form of such hidden motives, these simply reflect the political web in which war takes place and from which it cannot be isolated. War's instrumentality is not necessarily negated when the character of the policy changes or if true political motives are different to those publicly stated.[122] This web of politics is something within which policymakers and commanders are inevitably entwined. The cynical reasons for which force is often employed might degrade the purity of policy, but it remains policy nonetheless – warts and all – and will impact on military strategy, potentially detrimentally.

Implications

Limitations of policy control

As the previous section has shown, war is ensconced within and impacted on by the whole range of political conditions beyond the manipulation of actors: a substance 'removed from means and purpose that can be steered by the will'.[123] The primacy of policy has been overstated in the literature, with the result that some scholars believed Clausewitz's theory presented a doctrine which, if followed, would allow for the limitation of war: the use of force could be restricted 'to a scale that is no greater than necessary to achieve the objectives at stake'.[124] This interpretation has clouded Clausewitz's true meaning and has emphasised subordination over continuation.

121 See Colin S. Gray, *Another Bloody Century: Future Warfare* (London: Weidenfeld & Nicolson, 2005), p. 279.

122 See Gray, *Bloody*, p. 378.

123 Herberg-Rothe, *Puzzle*, p. 142.

124 Osgood quoted in Echevarria, *Clausewitz*, p. 87.

War is an instrument, but it is a notoriously 'unstable instrument'.[125] Clausewitz's choice of the term 'instrument' is thus unfortunate. As Echevarria notes, policy must understand that its instrument is a dynamic one: 'War involves living forces rather than static elements; thus, it can change quickly and significantly in ways the logic of policy may not expect'.[126] War can often generate effects entirely unintended by its users and is very difficult to keep within bounds. How much policy dominates in any particular war is a matter of circumstance. Policy is never tyrant: war has its own imperatives, requirements and dynamics. Also, Echevarria is correct to remind us that war's grammar 'places as many restrictions on policy as grammar does on language'.[127] In other words, policy must understand the limits of its instrument when determining objectives.

Dynamic feedback

The analysis so far has treated the relationship between politics and war in a predominantly static, linear fashion to elucidate the basic ideas. As Beyerchen notes, the usual approach 'envisions a compartmentalisation of politics and war in a linear sequence – first comes politics/policies, then war, then politics/policies again to make or maintain peace'.[128] The concept of subordination paints a picture of a one-way relationship whereby policy sets the goals for war to achieve and remains the chief consideration until the objective is either attained, lost or some marginal outcome is reached. This is not an image Clausewitz intended to convey and in fact there will be a constant, complex, dynamic and nonlinear feedback process between politics, policy and war. He notes that 'the political situation can change from year to year'[129] and, in an important sentence, that 'the original political objects can greatly alter during the course of the war and may finally change entirely since *they are influenced by events and their probable consequences*'.[130]

Because the major lines that run through war are political, military developments will have political effects which will impact on the decisions and calculations of political leaders. Thus, in war there is a constant interplay between the use of force, ends and means. The incredible complexity this encompasses means that, in theoretical terms, all we can hope to achieve is to draw attention to its prominent characteristics. The concept of 'continuation' reveals more strongly that the interaction of belligerents continually creates new political realities. Because new

125 Smith, *On Clausewitz*, p. 105.

126 Echevarria, *Clausewitz*, p. 94.

127 Echevarria, *Clausewitz*, p. 89.

128 Alan Beyerchen, 'Clausewitz, Nonlinearity and the Importance of Imagery'. Available online at: <http://www.clausewitz.com/readings/Beyerchen/BeyerschenNonlinearity2.pdf>, retrieved 20 August 2009.

129 *On War*, VIII, 4, p. 725.

130 *On War*, I, 2, p. 104. Emphasis added.

power relationships are created during the course of war, the interests of third parties, domestic factions and local populations might prompt them to intervene or react in myriad ways, with all the political and military consequences such moves entail. This is often the case when the military success of one group encourages others to support its cause, confident in the belief they are backing the winning side.

In this way, military developments impact on political dynamics and feed back into political decisions that, in turn, impact on war. It is important to remember that behind the abstract notion of 'policy' there are human decision-makers. Those responsible for setting policy do not disappear at the outbreak of hostilities, but rather will have to respond, adapt and reassess in the light of changed circumstances. Policy responds actively or passively to the effects of its instrument.

Ambiguity

Finally, Clausewitz believed the influence of policy on war was essentially ambiguous in relation to the rise to extremes: the political purpose neither necessarily causes a rise to extremes nor limits war, but it is the central, though not only, determinant of both. We should be careful not to mistake war's subordination to policy for the limitation of war, as some have been wont to do. When policy requires an extreme effort, war will follow precisely because of its element of subordination. He states that, the closer 'political probabilities drive war toward the absolute, the more the belligerent states are involved and drawn into its vortex'.[131] Clausewitz's dialectical argument in Book 1, Chapter 1 demonstrates that limited and unlimited wars are politically determined, whatever appearances may suggest.

But the matter is more complex because war creates its own dynamics. Wars begun for ostensibly limited objects, may escalate disproportionately to the original cause. War in reality is shaped by factors other than the political object – such as the effects of friction, uncertainty, passion and cultural norms – that cause the use of force to be modified. In wars where policy essentially becomes a struggle for existence, those extraneous forces inhibiting violence tend to be marginalised and only war's inherent friction might serve to limit its ferociousness.

Furthermore, policy is ambiguous in terms of its effect on strategic performance. The fact that Clausewitz believed war should be subject to the guiding intelligence of policy was not to argue that policy is always wise or a positive strategic force. For instance, policy can be unclear, make unrealistic demands on the military or may itself be wrong. On the other hand, sensible policy can have a significantly positive impact on strategic effectiveness – clearly stated objectives provide the military with a firm basis on which to plan operations.

---

131 *On War*, VIII, 3, p. 706.

*

Clausewitz believed that a rational thread derived from the overarching direction of policy runs through war, connecting individual combatants to the higher purpose of the war. The extent to which this influence is present depends on circumstances and the character of belligerents, but it is mistaken to conceive of war as ever being conducted without some guiding purpose, be it shallow or vaguely articulated. Groups attempt to achieve their ends through military means, even if this is a highly difficult undertaking given the difficulty of translation between the two realms, the uncertainty surrounding all decisions, the unstable nature of the military instrument and the unpredictable effects that the use of force entails. He did not believe war could always be employed in purely rational ways in pursuit of purely rational objectives. The relationship between policy and war is certainly not linear or narrowly deterministic, but rather reciprocal and dynamic. Even if policy is definite about what it intends to achieve, the grammar of war can cause objectives to radically change during hostilities and the extent to which policy remains in control is highly contingent.

Placing emphasis on the political nature of war focuses attention on the fundamental currents which underlie the use of organised force: the logic of the distribution of relative power between groups. Power may be manifested in radically different forms, but at heart it is about the ways in which social actors attempt to shape the context of their existence. When force is employed, the logic of politics does not cease; it continues in its most potent manifestation. Sometimes the main political lines in war will be distinct, at others they will be barely perceptible or shrouded by the brute clash of military forces. Often the political complexity can appear overwhelming, particularly when war is composed of myriad actors each pursuing divergent agendas, yet this simply reflects the reality of the complexity of political relationships. In war, the scales of political power are constantly weighed and re-weighed, victory is rarely final and military success can promptly be displaced by political failure.

War is an instrument of policy, but not only that. It is also a continuation of the web of political interactions carried on through the medium of force. This chapter has sought to underline the paradoxical impact of politics on war. On one level, policy provides war with a rational structure, whereby belligerents seek to attain their ends through the use of force. On the other hand, precisely because war is inherently political, it is embedded in a 'political web of war' and this explains war's objective unpredictability and non-rational aspects. The thread of reason may be only very thinly woven through the whole; the tapestry hangs together loosely.

War impacts on power constellations and dynamics at multiple levels and in endless ways which no actor can entirely control. Furthermore, extraneous political considerations encourage decisions which do not correspond directly with objectives or make strategic sense. Not only is pure rational action impossible, but attempts to achieve rational outcomes can be thwarted by the play of chance, irrational impulses, external political developments or even overzealously rational approaches. War is an instrument that can easily slip from the hands of those that wield it or, while solving certain problems, can create others where least expected. Those contemplating war as a means to achieve their ends must understand the nature of the instrument and attempt to read the political landscape in order to employ force in a way that produces desired effects while limiting adverse, unexpected developments.

Modern states have struggled to employ their militaries as effective instruments of policy, primarily due to complex political conditions rather than any significant military shortcomings. Particularly in situations of insurgency and terrorism – the pre-eminent forms of warfare faced by the West today – politics is central: as Paul Cornish demonstrates, counterinsurgency 'must be political first, political last, political always'.[132] Force can be an extremely blunt instrument in dealing with such forms of organised violence and great care is required to avert unintended consequences. This Clausewitz recognised in an era dominated by the annihilation battle – as exemplified in Napoleon's campaigns – and it is testament to his searching intellect that he saw beyond dominating impressions to comprehend that, in war, politics is 'the first, the supreme, the most far-reaching' consideration.[133]

132 Paul Cornish, 'The United States and Counterinsurgency: "Political First, Political Last, Political Always"', *International Affairs*, 85: 1 (January 2009), p. 79.

133 *On War*, I, 1, § 27, p. 100.

Chapter 5
Chance: The Realm of Uncertainty[1]

This chapter examines the element of Clausewitz's 'remarkable trinity' which concerns 'the play of chance and probability within which the creative spirit is free to roam'.[2] Clausewitz explains how war, like 'no other human activity is so continuously or universally bound up with chance' and that 'there is an interplay of possibilities, probabilities, good luck and bad that weaves its way throughout the length and breadth of the tapestry'.[3] Clausewitz's description may appear little more than his 'common-sense bordering on wit',[4] however this ignores the fact that others have overlooked its importance. Earlier thinkers had either marginalised its importance or subscribed so fully to its force that they abandoned all hope of producing concrete theory, while today its anticipated negation represents the basis of an entire school of thought which contends that a technological Revolution in Military Affairs (RMA) promises to dispel Clausewitz's 'fog of war'.[5] Colin Gray reminds us that, 'commonplace though emphasis upon the role of chance in war may appear, such emphasis is highly unusual among strategic theorists'.[6]

Humans like to feel in control of their environment, to be the masters of their fate and may feel uncomfortable when things have to be left to chance, the course of events cannot be reliably predicted or, as Machiavelli suggested, fortune appears as a fiendish Goddess wanting to be 'mistress of all human affairs'.[7] Such thinking can lead to misunderstanding, strategic myopia or even historical myth.[8] For such reasons, Clausewitz's analysis remains important today: it forces us to

---

1 This chapter uses material from: Thomas Waldman, '"Shadows of Uncertainty": Clausewitz's Timeless Analysis of Chance in War', *Defence Studies*, Vol. 10, No. 3, September 2010, pp. 336–68, reprinted by permission of Taylor and Francis Ltd (http://www.tandf.co.uk/journals).

2 Carl von Clausewitz, *On War*, trans. and ed. Michael Howard and Peter Paret (New York: Everyman's Library, 1993), I, 1 § 28, p. 101.

3 *On War*, I, 1, § 21, p. 97.

4 Marx quoted in Beatrice Heuser, *Reading Clausewitz* (London: Pimlico, 2002), p. 13.

5 The popular classic is Bill Owens, *Lifting the Fog of War* (Baltimore: Johns Hopkins University Press, 2001).

6 Colin S. Gray, *Modern Strategy* (Oxford: Oxford University Press, 1999), p. 95.

7 Quoted in Hanna F. Pitkin, *Fortune is a Woman* (Chicago: University of Chicago Press, 1999), p. 153.

8 National successes due more to exceptional luck may be written in terms of brilliant generalship. Michael Howard, *Causes of War* (Cambridge, Massachusetts: Harvard University Press, 1983), p. 190.

confront reality and to emphasise the demands chance and uncertainty places on human action, will, skill and creativity.

The chapter will begin by considering some of the influences, precedents and experiences that shaped Clausewitz's thought on the subject. It will then combine interpretation of Clausewitz's work with historical evidence and modern scholarship to develop a comprehensive conception of this tendency. Finally, it will consider some of the theoretical implications of the analysis.

Influences

The origins of Clausewitz's ideas on this subject are certainly somewhat unclear and diverse: aspects of his own experiences merge with imagery reflective of Romantic literature; humanistic philosophy finds its place next to concepts drawn from the physical sciences. Yet, despite this eclecticism, the following sections draw attention to three areas for which there is strong reason to believe they contributed to his ideas: precedents in existing military theory, Machiavelli and Clausewitz's own experiences.

Existing military theory

Clausewitz was adamant that earlier Enlightenment theorists had devoted insufficient attention to the concepts of chance and uncertainty. Rationalist theorists emphasised those aspects of war that were tangible, controllable and calculable. Such thinkers were not entirely blind to the existence of chance, rather it was the nature of their approach that precluded the incorporation of these factors into their systems: a form of 'methodological determinism'.[9] As Gat states, 'Enlightenment thinkers were quite aware of the factors of uncertainty but focused on what they considered to be suitable for intellectual formulation'.[10] Clausewitz explicitly attacked such views when he stated that those thinkers aimed at 'fixed values; but everything in war is uncertain and calculations have to be made with variable quantities'.[11] While some theorists appreciated the impact of chance, they failed to formulate this into clear concepts or develop them in theory. As Clausewitz noted, such factors were vital because they were ultimately what differentiated 'real war from war on paper' and which theory must tackle if it is to remain relevant.[12]

The Enlightenment theorists' failure to seriously analyse chance might be characterised as a form of theoretical denial or what today we might term

9 Jerome G. Manis and Bernard N. Meltzer, 'Chance in Human Affairs', *Sociological Theory*, Vol. 12, No. 1, March 1994, p. 53.

10 Azar Gat, *A History of Military Thought: From the Enlightenment to the Cold War* (Oxford: Oxford University Press, 2001), p. 187.

11 *On War*, II, 2, p. 156.

12 *On War*, I, 7, p. 138.

'cognitive dissonance'[13] – the refusal to contemplate that which you cannot reliably explain. Theory would only concern itself with subjects explicable through observable laws. Determined to reduce warfare to a system, they focused on those areas most susceptible to precise calculation. They started from 'the bottom up', developing systems concerned with tactical matters or 'fields where the enemy's independent will did not have to be taken into account', rather than with more complex and intangible problems associated with strategy.[14] Foremost amongst these was the science of siege warfare, which encouraged the belief that theory could confine itself to mathematical prescriptions that would leave almost nothing to chance.[15]

Later theorists such as Guibert and Bülow, although army officers by training, had little experience of actual warfare and were thus poorly qualified to appreciate the effect of chronic uncertainty. This is reflected in Bülow's concentration on geometrical lines of operation, whereby elaborate strategic planning would serve to reduce 'the element of pure chance that Frederick [the Great] had feared in it'. Armed with such principles he believed planning could become 'more fruitful, prediction somewhat more possible, warfare more of a 'science'.[16]

It would be inaccurate to claim that previous thinkers had not recognised the difficulties that chance presented for theory. Maurice de Saxe recognised that often outcomes will be 'dependent upon the favor of fortune, which sometimes is very inconstant'[17] and that, as a result, commanders require courage and intelligence, 'a talent for sudden and appropriate improvisation' and the ability 'to see the opportunity and to know how to use it'.[18] Likewise, Jomini recognised uncertainty as 'one of the chief causes of the great difference between the theory and the practice of war'.[19]

Yet, as Echevarria notes, the problem is not so much that Jomini assumed a world without friction, but rather that his underlying assumption was 'that if a commander were to hold fast to [his] proposed principles, and use them as guidelines, he would succeed in the face of the confusion and chaos of battle'.[20] Furthermore, Jomini makes the point that if success was achieved by chance, then it would be chance action in line with principles, nothing else. Clausewitz

---

13 Ken Booth, *Strategy and Ethnocentrism* (London: Croom Helm, 1979), p. 25.

14 Martin van Creveld, *The Art of War: War and Military Thought* (New York: Smithsonian Books, 2005), p. 78.

15 Clausewitz acknowledged that siege warfare was primarily a matter of mathematics and geometry. *On War*, III, 15, p. 251.

16 Palmer, 'Dynastic to National War', p. 117.

17 Maurice de Saxe, *Reveries on the Art of War*, trans. and ed. Thomas R. Phillips (New York: Dover Publications, 2007), p. 18.

18 Saxe, *Reveries*, pp. 117–19.

19 Antoine Henri de Jomini, *The Art of War*, trans. G. H. Mendell and W. P. Craighill (Texas: El Paso Norte Press, 2005), p. 215.

20 Antulio J. Echevarria II, *Clausewitz and Contemporary War* (Oxford: Oxford University Press, 2007), p. 108.

believed theorists like Jomini directed their principles towards 'unilateral activity' and considered only factors that 'could be mathematically calculated'. They thus formulated wholly 'lopsided principle[s] that could never govern a real situation'.[21]

Saxe and Jomini failed to grasp or sufficiently articulate that chance could not simply be viewed as something incidental to theory, but was in fact central to it. As Daniel Moran notes, even those fully aware of war's uncertain nature,

> might feel that the surest path to clarity requires that incidental difficulties be ignored, in the same way that a scientist seeking a constant pattern or signal within a mass of data is entitled, indeed required, to ignore the 'noise' that surrounds it. For Clausewitz it was unrealistic to adopt such an attitude to war, in which the effects of chance are so profound that they become the signal, the central reality, and not an exogenous variable to be discounted.[22]

Given these theorists' tendency to marginalise chance, it is not surprising that detailed discussions of genius are largely absent or mentioned only in passing. They devoted little attention to this subject or situated it in a 'sublime' realm presumed 'beyond' theory. If war is determined by rational laws, what role is there for genius when presumably the enemy will submit following the application of logical principles? Such wars would be decided by superior algebra. This led to a curious position, which Clausewitz would later mock, whereby anything not clearly explicable through 'correct principles' must be due to the workings of sublime genius which is able to spurn the rules. As Echevarria notes, 'They regarded genius as a rare, inexplicable, and therefore inconvenient phenomenon'.[23]

However, at the other extreme, the Romantic world-view depicted war as 'the sphere of clashing wills, rising emotions, uncertainty, and confusion … diverse complexity and endless confusion; [that] could only be mastered by the general's practical genius and iron will'.[24] The *Sturm und Drang* writers emphasised the play of chance in their works, as well as the creative will that must overcome it: a central theme in Schiller's *Wallenstein*, which was one of Clausewitz's favourite works.[25] That Clausewitz's own inquiries into the psychological qualities of great commanders often stressed character and spirit over education or cognitive ability, betrays a debt to the Romantics.[26]

---

21 *On War*, II, 2, pp. 155–6.

22 Daniel Moran, 'Moran, Daniel, 'Strategic Theory and the History of War', Naval Postgraduate School. Available online at: <http://www.clausewitz.com/readings/Moran-StrategicTheory.pdf#zoom=100>, retrieved 20 January 2012, p. 8.

23 Echevarria, *Clausewitz*, p. 102.

24 Gat, *Military Thought*, p. 315.

25 Gat, *Military Thought*, pp. 181–2.

26 Gat, *Military Thought*, p. 183.

In this vein, Berenhorst's work articulated a view of war determined purely by the contingent, exceptional and unpredictable.[27] He held that, 'far from being scientific, war was anarchic, dominated by accident ... Efforts to control, let alone abolish, this primeval wildness were absurd'.[28] Clausewitz was at least sympathetic to the causes of Berenhorst's despair, yet would not follow him in his submission to fortune. Clausewitz's conception of genius differed from that of the Romantics, so where 'Goethe's genius acted spontaneously ... Clausewitz's genius took action only after correctly assessing the overall situation'.[29] His was an altogether more cerebral genius. The teaching of Scharnhorst was vital here. Even though the Berlin Academy was still teaching the Enlightenment dogma that sound theory could eliminate chance, Scharnhorst was encouraging his students to consider the more realistic notion of 'the ability of theory to help men deal with surprise, to help them exploit the unforeseen'.[30]

So, existing theory either avoided the subject or, where chance was considered, it was marked by an excessively melodramatic reaction. Nevertheless, some of the ideas that Clausewitz would ultimately develop are detectable. Moving beyond these precedents, Clausewitz would develop a conception which represented neither a capitulation to fortune nor an overoptimistic belief in theory's ability to overcome it. It appears Clausewitz had to reach further back in time to find a thinker – Machiavelli – who had appreciated more fully the importance of chance in war.

Fortune and virtue in Machiavelli

We know that Clausewitz was an avid reader of Machiavelli and believed that 'his judgement in military matters is very sound'.[31] Machiavelli's rendering of *fortuna* and *virtù* – which pervade all his major works[32] – furnished Clausewitz with the conceptual keys to his analysis. *Fortuna* for Machiavelli held a similar meaning to the 'chance' employed by Clausewitz, in essence referring to the incalculable and the fortuitous. Machiavelli held that, 'The continued existence of struggles and uncertainties patterns the character and the methods of war: there is no safe course'.[33] Neal Wood notes that for him, 'In no other human situation are chance,

27 Michael Howard, *Clausewitz: A Very Short Introduction* (Oxford: Oxford University Press, 2002), p. 23.

28 Peter Paret, *Clausewitz and the State: The Man, His Theories, and His Times* (Princeton: Princeton University Press, 1985), p. 202.

29 Echevarria, *Clausewitz*, p. 112.

30 Paret, *Clausewitz*, p. 71.

31 Paret, *Clausewitz*, p. 169.

32 Niccolò Machiavelli, *The Discourses*, trans. Leslie J. Walker (London: Penguin, 2003); *The Prince*, trans. George Bull (London: Penguin, 1999); and *The Art of War*, trans. Ellis Farneworth (Cambridge, Massachusetts: Da Capo Press, 2001).

33 Gilbert, 'Machiavelli: The Renaissance of the Art of War', in Paret, *Makers*, p. 24.

accident, and uncertainty … so prominent as in the peril and hardship of military encounter'.[34]

In the penultimate chapter of *The Prince*, Machiavelli notes how many people in his own times, because of 'the great variations and changes, beyond human imaginings', had come to believe that, 'there is no point sweating over things, but that one should submit to the rulings of chance'. Machiavelli does not subscribe to this fatalistic view, but instead proposes a compromise that allows for the operation of human will: 'fortune is probably arbiter of half the things we do, leaving the other half or so to be controlled by ourselves'.[35] He did not share the widespread contemporary belief that man is entirely in the hands of *Fortuna*.[36]

This fusing of chance and human will broadly reflects the duality that Clausewitz proposes, for instance when he states that, 'with uncertainty in one scale, courage and self-confidence must be thrown into the other to correct the balance'.[37] Machiavelli explains how fortune is not a wholly overpowering force; like a violent raging river, one can still take precautions to control its power.[38] Here, both share the belief that in facing the vicissitudes of chance, men must adapt to changing times because, 'fortune is changeable whereas men are obstinate in their ways, men prosper so long as fortune and policy are in accord, and where there is a clash they fail'. Machiavelli declares that 'fortune is a woman' and as such one can only 'command her with greater audacity'.[39]

The key for Machiavelli is acting when fortune favours you – what we might term opportunism. As he states, 'Nothing is of greater importance in time of war than knowing how to make the best of a fair opportunity when it is offered'.[40] In *The Prince* he explains that outstanding leaders 'do not seem to have from fortune anything other than opportunity. Fortune, as it were, provided the matter but they gave it its form; without opportunity their prowess would have been extinguished, and without such prowess the opportunity would have come in vain'.[41] Machiavelli's insights helped Clausewitz develop a conception of chance which was, if not unambiguously positive, at least malleable in the hands of great ability.

The qualities that enable men to face fortune are encapsulated in Machiavelli's concept of *virtù*, which strikingly parallels Clausewitz's discussion of genius. These similarities are noted by Wood, who states that Clausewitz's 'analysis of the qualities of generalship and of military virtue can be interpreted as an illuminating

34 Neal Wood, 'Seneca and Machiavelli', p. 15 and p. 20.
35 Machiavelli, *The Prince*, p. 79.
36 Gilbert, 'Machiavelli', p. 24.
37 *On War*, I, 1, § 22, p. 97.
38 Machiavelli, *The Prince*, p. 80.
39 Machiavelli, *The Prince*, p. 82.
40 Machiavelli, *The Art of War*, p. 202.
41 Machiavelli, *The Prince*, p. 18.

commentary on Machiavelli's concept of *virtù*'.[42] The attributes associated with *virtù* in Machiavelli's work are distinctly Clausewitzian: 'boldness, bravery, resolution, and decisiveness' combined with 'endurance and firmness, the necessary resilience, [and] the power of sustaining a course of action until the end is achieved'.[43] *Virtù* is associated with 'vital creative energy'[44] and resembles the notion of 'creative spirit' in the trinity. *Virtù* for Machiavelli and Clausewitz is never simply the Romantic notion of 'unruly energy, unbridled ferocity, and rapidly exhausted boldness', rather intellect and knowledge serve to direct and control the bold creative urge.[45]

So, for both, chance is not necessarily cause for despair, but is nevertheless a powerful force that cannot be avoided. As Paret notes of Clausewitz, 'to exclude or deny chance was to go against nature … Despite its constant power, chance was more than danger: it was a positive force to be exploited'.[46] Fortune rules mostly where men lack *virtù* or genius. These attributes apply to all actors in war: Machiavelli's *virtù* was not confined to the commander, but was the quality that enabled any human to confront the whims of fortune. Clausewitz equally emphasised the 'virtues' that enabled ordinary soldiers to face the possibility of death.

Clausewitz's experiences

Clausewitz believed that the best proofs in theory are personal and historical experience: careful consideration of both emphasised to him the enormous distance that separated actual warfare and existing theories. They presented precise and sterile principles which could not be reconciled with the confused and messy reality he had witnessed. What was the element that kept theory so divorced from reality? Early in his studies, he recognised that the missing factor was the uncertainty, unpredictability and friction of war. Following the Wars of the First Coalition, his regiment was endlessly drilled in antiquated methods and Clausewitz would later recall his disdain for the 'formalistic, ceremonial character of these manoeuvres'[47] when, 'even a modicum of reflection on these exercises … was bound to lead at once to the realisation that none of this had taken place in the war that we had fought'.[48]

Clausewitz was struck by the omnipresence of chance events in the wars in which he had fought; perhaps none more so than during the disastrous twin

42 Neal Wood, 'Introduction', in Machiavelli, *The Art of War*, p. xlvi.

43 Machiavelli, *The Art of War*, p. lv.

44 Machiavelli, *The Art of War*, p. lv.

45 Machiavelli, *The Art of War*, p. lv.

46 Paret, *Clausewitz*, p. 203.

47 Paret, *Clausewitz*, p. 45.

48 Carl von Clausewitz, 'Observations on Prussia in Her Great Catastrophe', in Clausewitz, *Writings*, p. 40.

defeats at Jena-Auerstedt in 1806. The Prussian campaign, Clausewitz observed, was dogged by every kind of friction from divided command, indecision, insubordination, disorder and confusion.[49] The defeat was also partly caused by the failure of Prussian intelligence.[50] Clausewitz's dismissive views of intelligence may well derive from this experience and Napoleon's skilful use of deception and operations security may have persuaded him of the difficulty of ever accurately determining enemy intentions. So, Clausewitz concluded his lectures on 'little war' in 1811 with the following modest statement:

> As little combat experience as I have, it is enough to give me an accurate view of the way most episodes in war unfold, as well as the numerous chance incidents, which touch everything, and of the numerous difficulties that inhibit accurate execution of the precise plans that theory tends to formulate … [They] can be recognised only through experience.[51]

Perhaps also Clausewitz was compelled to emphasise this aspect of war due to the extent to which the character of the wars of the Napoleonic era were more suffused with chance than those of the eighteenth century, and even those of the First Coalition, as 'rules and geometric calculation gave way to passion, massive friction and uncertainty'.[52] The French Revolution made possible mass conscription, which necessitated far-reaching changes in the organisation, supply, discipline and training of the new armies, which in turn would greatly increase the 'tempo and range of operations' – the wars of this period would become 'more complex and less predictable' than before.[53]

These personal experiences, combined with his detailed analyses of various campaigns revealed how outcomes were frequently decided by chance and luck, how the detailed plans of commanders rarely survived contact with the enemy and how mistaken reports on the strength or dispositions of the enemy often were. The challenge, however, would be explaining such factors in theory, developing coherent concepts and revealing their relationship to other aspects of war.

Explorations

Now we have some idea of the various influences on Clausewitz's thought, we can explore the subject in greater detail. The various concepts of chance, uncertainty, probability, friction and the fog of war are all strongly related and interconnected. Nevertheless, Clausewitz meant something specific in his use of these individual

49 Clausewitz, 'Observations', pp. 32–84.

50 Hugh Smith, *On Clausewitz* (Hampshire: Palgrave Macmillan, 2005), p. 150.

51 Quoted in Paret, *Clausewitz*, p. 191.

52 Parkinson, *Clausewitz*, p. 312.

53 Paret, *Understanding*, p. 78

concepts, therefore it is important to explore the distinguishing features of each. Equally, the various concepts clearly constitute a coherent 'family' of ideas connected by a common thread. Modern interpreters often emphasise the integration of these concepts, employing generic terms such as 'general friction', which encompasses danger, exertion, uncertainty and chance.[54] Such approaches are welcome, but only if based on a solid grasp of the individual concepts that comprise them.

The intruder: chance

The concept of chance employed by Clausewitz is intended in its pure form, as 'arbitrary and incalculable' events characterised by an absence of predictability or regularity in the sequence of action and its antecedent conditions.[55] Chance is the 'intruder' that 'makes everything uncertain and interferes with the whole course of events'.[56] If chance renders war the 'domain of the unexpected',[57] one can perhaps understand the 'cognitive dissonance' displayed by Enlightenment theorists given that it represents the negation of the very uniformity they sought to uncover.[58]

Because the very nature of something that happened by chance is that it was unforeseen, we can prepare only in the general sense of anticipating its probable existence; as something we know will occur, but we are ignorant as to the precise form it will take. It was recognition of this fact that compelled the elder Moltke to claim that there was no point in devising detailed schemes beyond the deployment phase.[59] Clausewitz explains how war is 'rich in unique episodes. Each is an uncharted sea, full of reefs. The commander may suspect the reefs' existence without ever having seen them; now he has to steer past them in the dark'.[60] In war, success or failure rarely accrues simply from the skill or incompetence of the commander, but is greatly determined by good luck or misfortune. This is not something humans like to accept: Clausewitz describes the proclivity to explain success as the result of genius alone as a 'gratifying assumption'.[61]

In addition to this standard conception, Alan Beyerchen, drawing on the work of Poincaré, has identified two other principal forms of chance in Clausewitz's work.

---

54 See Barry D. Watts, *Clausewitzian Friction and Future War* (Washington: National Defense University, 1996).

55 Bassford, 'The Primacy of Policy and the "Trinity" in Clausewitz's Mature Thought', in Hew Strachan and Andreas Herberg-Rothe (eds), *Clausewitz in the Twenty-First Century* (Oxford: Oxford University Press, 2007), p. 89.

56 *On War*, I, 3, p. 117.

57 *On War*, I, 3, p. 119.

58 Manis and Meltzer, 'Chance in Human Affairs', p. 46.

59 Gat, *Military Thought*, p. 336.

60 *On War*, I, 7, p. 139.

61 *On War*, II, 5, p. 195.

Beyerchen refers to a form of chance deriving from 'analytical blindness'.[62] This is perhaps best conceived as chance that results from human mental weakness, or 'our frailty and ignorance' – an inability to comprehend the whole so that 'we may happen to overlook circumstances which, at first sight, seemed completely foreign to the anticipated fact, to which we should never have dreamed of attributing any influence'.[63] When things do not proceed as expected due to causes that lay beyond our mental horizon, such occurrences appear as blind chance – as Carr had it, they are 'the measure of our ignorance'.[64]

A large part of the experience of chance may in fact turn out to be due to this fundamental human weakness. No human mind is capable of comprehending simultaneously the complex whole, its interactions and the infinite relations between causes and effects. With hindsight, we may be able to better understand the circumstances that led to certain events, but the point remains that, for those involved at the time, whether due to sheer incompetence or analytical blindness, events were not foreseen. We must be mindful, as Clausewitz counselled, not to project our knowledge of what we know happened, back on to those who had to make decisions at the time.[65]

Another form of chance Beyerchen highlights concerns the concept of 'micro-cause' or what historians might know as the problem of 'Cleopatra's nose'.[66] Micro-cause refers to situations in which disproportionately large effects result from apparently trifling initial causes. Clausewitz draws attention to this problem in relation to his discussion of the critical analysis of past campaigns: the process is made difficult by the fact that 'in war, as in life generally, all parts of the whole are interconnected and thus the effects produced, however small their cause, must influence all subsequent military operations and modify their final outcome to some degree'. Furthermore, if such causes 'happen to be transitory or accidental, history may not have recorded them at all'.[67] The great gap between infinitely small and 'inappreciable' causes and such potentially enormous, inescapable effects means we have to ascribe to chance that which we could not hope to explain. This is not so much a function of human weakness as the objective fact of the incredible complexity of reality – Clausewitz emphasises the 'vast, almost infinite distance there can be between cause and effect'.[68]

---

62 Alan Beyerchen, 'Clausewitz, Nonlinearity and the Importance of Imagery'. Available online at: <http://www.clausewitz.com/readings/Beyerchen/BeyerschenNonlinearity2.pdf>, retrieved 20 April 2007.

63 Poincare, *Science and Method*, p. 75.

64 Edward Hallett Carr, *What is History?* (London: Penguin, 1990), p. 102.

65 *On War*, II, 5, p. 193.

66 This holds that it was the beauty of Cleopatra's nose that determined the outcome of the decisive Battle of Actium due to Antony's infatuation. See Carr, *What is History?*, pp. 87–108.

67 *On War*, II, 5, pp. 182–5.

68 *On War*, VIII, 1, p. 698.

To subscribe to the pervasive existence of chance in war is not to deny that things are causally determined – as Carr notes, accident is not an absence of causality.[69] The point is that causes in war are sometimes so incredibly small as to escape our notice and appear detached from observable effects or appear so unexpectedly from beyond our conceptual radar that they appear as chance. Put differently, we can maintain that all chance events have causes, they were just causes we failed to see or predict. They are nevertheless rightly considered objective, as far as they are both a consequence of the human condition and the incredible randomness of reality.

A final point to note is to be suspicious of commanders who retrospectively explain their own failure in war by an appeal to chance. Defeated generals will look to explain away their shortcomings by claiming the Goddess of Fortune was against them, when in success that same general would no doubt emphasise his complete control of the situation throughout.[70] Chance contributes to both success and failure, but rarely is it the sole arbiter of either: both talent and incompetence figure greatly. One might suspect an underlying ineptitude in those who claim all is decided by chance and a measure of undeserved self-assurance in those who deny its existence. In what was perhaps a rare instance of modesty, Napoleon was quoted as saying, 'Engage the enemy, and see what happens'[71] – it is telling that one of history's greatest commanders should pay heed to luck in this way.

The 'shadows of uncertainty'

Clausewitz remarked that 'war is the realm of uncertainty; three quarters of the factors on which action in war is based are wrapped in a fog of greater or lesser intensity'.[72] So, how are we to distinguish this concept from that of chance? Certainly these concepts are related in the sense that a psychological state of uncertainty is often a consequence of the play of chance. Yet, as Herbig notes, 'one may feel uncertain for many reasons other than chance, and chance does not always lead to uncertainty'.[73]

The concept of uncertainty essentially refers to the human reaction to that which cannot be fully known or controlled and concerns the psychological condition of those involved in war. As such, uncertainty becomes evident as integral to the nature of war as soon as we consider the human forces that are central to it. To ensure theory does not lose touch with reality, Clausewitz stresses that it 'must also take the human factor into account … The art of war deals with living and

---

69 Carr, *What is History?*, p. 99.

70 Hitler denied his own responsibility for failure as he dictated to his secretary Martin Borman in Spring 1945. Richard Overy, *Why the Allies Won* (London: Pimlico, 1996), p. 315.

71 Quoted in Herbig, 'Chance and Uncertainty', p. 108.

72 *On War*, I, 3, p. 117.

73 Herbig, 'Chance and Uncertainty', p. 104.

moral forces. Consequently, it cannot attain the absolute, or certainty; it must always leave a margin for uncertainty'.[74] War is not only about factors that can be counted or calculated, but is suffused with psychological factors which are inseparable from physical factors: unlike an alloy, they cannot be separated by chemical processes.[75]

All decisions in war must be taken with regard to consequences in this moral realm; whether this relates to one's own forces, government and wider society or those of the enemy and neutral actors. As Clausewitz states, it would be platitudinous to list all such phenomena because most are commonly known (such as the enemy's intentions and strength of will, public opinion, morale and so on). It is because these factors are inherently intangible and unpredictable – 'they cannot be classified or counted' – that attaining certainty as to their precise character or effect is necessarily limited regardless of the quantity or quality of information one might have.[76] Clausewitz notes how, for the commander, 'Thousands of wrong turns running in all directions tempt his perceptions; and if the range, confusion and complexity of the issues are not enough to overwhelm him the dangers and responsibilities are'.[77]

The extent to which uncertainty and confusion is typical in war is exacerbated by the incredible 'prevalence and potency of the factor of surprise'.[78] The enemy will be constantly trying to achieve the unforeseen arrival of their forces where not expected through the use of deception, secrecy, speed and cunning. What Clausewitz describes as the 'universal urge to surprise' can have a serious paralysing psychological effect on morale and 'loosen the bonds of cohesion' within the opposing force. Importantly, surprise, as Clausewitz emphasises, is often dependent on chance, coincidence and the mistakes or misperceptions of the opponent.[79] Surprise, therefore, stems from the weakness of human perception and basic uncertainties.

So, in gauging and evaluating such factors, Clausewitz extolled the virtues of intuition, the 'inward eye'[80] and 'discreet judgement'.[81] For instance, when discussing decisions regarding the levels of effort to be made and force to be employed, he states that such intangibles 'introduce uncertainties that make it difficult to gauge the amount of resistance to be faced and, in consequence, the means required and the objectives to be set', so the most appropriate approach is to,

74 I, 1, § 22, p. 97.

75 *On War*, III, 3, p. 216.

76 *On War*, III, 3, pp. 216–17.

77 *On War*, VII, 22, p. 693.

78 Colin S. Gray, 'How Has War Changed Since the End of the Cold War?', *Parameters*, Spring 2005, p. 15.

79 *On War*, III, 9, pp. 235–9.

80 *On War*, I, 3, p. 118.

81 *On War*, VII, 22, p. 693.

> gauge the strength and situation of the opposing state … gauge the character and abilities of its government and people and do the same in regard to our own. Finally, we must evaluate the political sympathies of other states and the effect that war may have on them.[82]

The stress on judgement over calculation is a direct consequence of uncertainty deriving from the inescapable presence of intangibles. This point is easily overlooked. Even for militaries equipped with the most advanced intelligence gathering technology, definitively determining potential effects and reactions in the sphere of the mind and spirit will remain an unattainable chimera. As David Kahn notes, 'not even the most energetic intelligence operation can penetrate an enemy's brain'.[83] Moreover, even the enemy might not fully 'know itself the limits of its ambitions'.[84]

Lack of proper understanding is often exacerbated by forms of ethnocentrism and strategic-cultural myopia, which distort accurate assessments of one's enemies' (and even one's own or allies') intentions and capabilities or the political environment. As Booth argues, this 'lack of empathy has meant an absence of intimate understanding of the feelings, thoughts and motives of others: this has prevented an accurate forecasting of likely responses'.[85] Clausewitz was well aware of the way such perspectives distort evaluations, which are not only a problem of obtaining objective knowledge, but also crucially dependent on the 'qualities of mind and character of the men making the decision'.[86] All actors make sense of the world through unique cultural perspectives. They observe, orientate themselves and act based on assumptions about the world. It is also easily forgotten that all information (even if derived from advanced sources) only attains meaning once it is digested by human agents and subsequently used as the basis for decisions – decisions that will inevitably be shaped by the unique outlook of the actor. The consequences arising from ethnocentric perspectives can be serious, exacerbate uncertainty and be highly detrimental to operations: for instance, a failure to understand enemy intentions can increase the likelihood of political, strategic or tactical surprise.[87]

The 'fog of war'

Regardless of the difficulties of mitigating uncertainty associated with moral forces, the information on which assessments are made in war is, for Clausewitz, ineluctably deficient. Clausewitz is well known for his emphasis on the

---

82 *On War*, VIII, 3B, pp. 707–8.

83 David Kahn, 'Clausewitz and Intelligence', in Handel, *Modern Strategy*, p. 119.

84 Booth, *Strategy*, p. 123.

85 Booth, *Strategy*, p. 40.

86 *On War*, VIII, 3B, p. 708.

87 Booth, *Strategy*, pp. 104–5. Consider, for example, Japan's misperception of America's expected reaction to the attack on Pearl Harbour.

interminable poverty of reliable information in war and his dismissive attitude regarding the substance and value of intelligence. In a notably evocative passage Clausewitz states that,

> the general unreliability of all information presents a special problem in war: all action takes place, so to speak, in a kind of twilight, which, like fog or moonlight, often seems to make things seem grotesque and larger than they really are. Whatever is hidden from full view in this feeble light has to be guessed at by talent, or simply left to chance.[88]

This general idea is expressed in many other parts of his work. Clausewitz states that, 'If we consider the actual basis of this information, how unreliable and transient it is, we soon realise that war is a flimsy structure that can easily collapse and bury us in ruins'.[89] At best, and even this may be doubted, Clausewitz notes that the 'only situation a commander can know is his own; his opponent's he can know only from unreliable intelligence'.[90] Moreover, it is likely that the enemy will be doing his level best to deny us knowledge of its intentions or even actively attempting to deceive, fool and surprise us. Also, intelligence is often false and inaccurate because fear causes people to exaggerate bad news.[91] Ken Booth has remarked that, 'Worst-case forecasting is to strategic analysis what the 'god of the gaps' is to theology: it fills in for what we do not understand'. In conditions of danger, where survival may be in jeopardy, there may be a proclivity for 'the possible to become thought of as the probable'.[92]

So, fear, intermixed with chance, numerous imponderables, the limitations of accurate observation and the vast amount of variables involved – before which 'Newton himself would quail' – are provided as some of the chief reasons for Clausewitz's scepticism with regard to the reliability of information.[93] Together, these factors create what Clausewitz memorably termed the 'fog of war'. However, as Kahn rightly points out, despite this intense scepticism regarding intelligence, Clausewitz does not 'dogmatically maintain that it can never serve'.[94] Clausewitz takes it for granted that more and better information will be sought, the potential value of which would be dependent on the commanders ability to effectively exploit what was made available, but more importantly, to be able to withstand the endless torrent of false or contradictory reports.[95] Nevertheless, the central

---

88 *On War*, II, 2, p. 161.

89 *On War*, I, 6, p. 136.

90 *On War*, I, 1, § 18, p. 95.

91 Such as overblown estimates of enemy strength. *On War*, I, 6, p. 136.

92 Booth, *Strategy*, pp. 122–6.

93 *On War*, VIII, 3B, p. 708.

94 Kahn, 'Intelligence', p. 117.

95 As Clausewitz states, 'The dangers that are reported may soon, like waves, subside; but like waves they keep recurring, without apparent reason. The commander must trust his

point for Clausewitz is that a superlative degree of accurate information is both an objective impossibility and a dangerously deceptive fantasy.

Of course, since Clausewitz wrote, intelligence gathering techniques have been enhanced by enormous strides in technology and institutional specialisation. Both the potential quality and quantity of information available to politicians, commanders and troops is vastly superior. Some of the major changes would include the development of large staff bureaucracies and permanent, professional intelligence bodies, signals intelligence and the more advanced cryptographical techniques associated with World War II. In more recent years, intelligence gathering techniques have become enormously technologically sophisticated. Yet, as argued by a number of specialist scholars, regardless of technological advances, failures of intelligence are inevitable. As soon as one explores the processes and organisational dynamics of intelligence acquisition and analysis, the potential for failure becomes clear. The sources of this 'inevitability of failure' – particularly at strategic levels – have been masterfully studied by Richard Betts, Robert Jervis and Michael Handel.[96]

These authors point to such factors as basic human psychology, ingrained preconceptions, the nature of the relationships between consumers and producers of intelligence, differences of opinion amongst analysts and decision-makers, but perhaps most importantly the 'wishful thinking, cavalier disregard of professional analysts and … the premises and preconceptions of policy makers'.[97] Also, one only has to factor in the enemy's almost certain efforts in the realm of counter-intelligence, misinformation, deception and so on – none of which can be definitively negated – and scepticism creeps back in. Knox notes that 'intelligence collection devices obey the same dialectical law as weaponry. Counteraction or evasion or deception may at any point surprise even – or especially – those made slothful by seemingly effortless technological superiority'.[98]

While perhaps Clausewitz was excessively dismissive of intelligence given the limited nature of the art in his time, it would be just as mistaken to be overly impressed by the claims of those 'information warriors' who stalk the corridors of modern military organisations. As Keegan states, 'there is no such thing as the golden secret, the piece of "pure intelligence", which will resolve all'.[99] Recent

---

judgement and stand like a rock on which the waves break in vain'. *On War*, I, 6, p. 136.

96 Richard K. Betts, 'Analysis, War, and Decision: Why Intelligence Failures Are Inevitable', *World Politics*, Vol. 31, No. 1, October 1978, pp. 61–89; Robert Jervis, 'Hypotheses on Misperception', *World Politics*, Vol. 20, No. 3, April 1968, pp. 454–79; and Michael Handel, *War, Strategy and Intelligence* (London: Routledge, 1989), pp. 187–281.

97 Betts, 'Analysis', p. 67.

98 Knox, 'Conclusion: Continuity and Revolution in the Making of Strategy', in Williamson Murray, MacGregor Knox, and Alvin Bernstein (eds), *The Making of Strategy: Rulers, States and War* (Cambridge: Cambridge University Press, 2007), p. 642.

99 Keegan, *Intelligence*, p. 4.

high-profile Western intelligence failures – in relation to the 9/11 attacks and Iraq's putative WMD arsenal – only add credence to these insights.

Lifting the fog?

In recent decades a so-called Revolution in Military Affairs (RMA) has been proclaimed. Its more extreme proponents believe it will lift the 'fog of war' and remove friction and uncertainty. This idea is centred around the impact of new technologies and took hold seriously in the 1990s. Its material components can be divided into a number of 'systems' including sensors, communications and advanced weaponry. However, it is their complete integration – the 'system of systems' – that represents what is specifically novel. Associated doctrine focuses on the prospect of achieving 'dominant battlespace knowledge' and holds out the potential for the seamless collection, assessment and communication of information, allowing force to be applied through advanced weapons systems with greater range, lethality and accuracy. All this contributes to enhanced situational awareness and improved mission effectiveness, or 'Full Spectrum Dominance'.

Yet, despite these far-reaching developments, many commentators have emphasised the limitations of such claims. Indeed, many of the problems Clausewitz identified remain: inevitably, new solutions spawn new problems and new dependencies create new vulnerabilities.[100] For instance, there is a serious potential problem of information overload, whereby commanders cannot comprehend the immense 'noise' they are faced with: as Clausewitz put it, 'we know more, but this makes us more, not less, uncertain'.[101] Once all information has been collected it still has to be comprehended, interpreted and acted on. At all stages, human fallibility and innumerable complications intervene, reflecting the eternal problem of converting information into knowledge and knowledge into action. Overemphasis on new technologies can also lead to a form of information dependence, distracting attention away from crucial intangibles such as morale, the orientation of allies or the state of domestic opinion: such 'strategic tunnel vision' potentially downplays the value of human perception of cultural and political dynamics.[102]

The deepest cause for scepticism relates to the challenge of converting technological superiority into political effect. If indeed a revolution is taking place, it is principally occurring at tactical/operational levels, whereas in strategy success is not so narrowly dependent on better weaponry and equipment. Lawrence Freedman notes that, 'the real difficulty is that military power can only be truly

100 Colin S. Gray, *War, Peace, and Victory: Strategy and Statecraft for the Next Century* (New York: Simon and Schuster, 1990), p. 111.

101 *On War*, I, 3, p. 117.

102 Colin Gray, 'The RMA and Intervention: A Sceptical View', in Colin McInnes and Nicholas J. Wheeler (eds), *Dimensions in Western Military Intervention* (London: Frank Cass, 2002), p. 55.

judged against the political purposes it is intended to serve'.[103] Advanced systems may enable rapid victories against conventional forces, however this by no means guarantees their translation into meaningful political outcomes. This is not to argue that information superiority is not worth attaining, only that it should not be seen as a 'magical elixir', particularly at the strategic level. So, should Clausewitz's scepticism be viewed as hopelessly time-bound? The weight of professional opinion would suggest not. As Gray succinctly concludes, 'no technical panacea can eliminate uncertainty altogether'.[104]

Decision-making dynamics

The paucity of reliable information is further exacerbated by clashes of opinion amongst decision-makers, precisely because the lack of information inevitably leaves room for diverse judgements. Clausewitz explains how such disputes are often overlooked retrospectively because they are 'seen as the scaffolding that can be removed once the building is complete'.[105] In other words, when we look back on a successful operation, for instance, it is rarely the difficult decisions that led to action that are the focus, but rather the action itself and its outcome – this has the effect of downplaying the fundamental problem of *choice* that decision-makers faced, so that 'what the layman gets to know of military events is usually nondescript… [and] it would be impossible to guess what obstacles were faced and overcome'.[106] Clausewitz could not fail to recognise this problem as such paralysis in decision-making was largely to blame for Prussia's defeat in 1806. Probably with such events in mind, he states that, 'Nowhere are differences of opinion so acute as in war, and fresh opinions never cease to batter at one's convictions'.[107]

In this respect, if there was ever a suitable subject for counter-factual reasoning, then strategic history is a prime contender. 'What if' questions encourage consideration of the acute problems actors faced, often lost in narrative accounts whereby events are explained with an underlying air of inevitability. Of course, things did turn out as they did for good reason and thus lingering too long in this realm of 'might-have-beens' can be unhealthy – for things to have been different, the causes would also have had to be different[108] – nevertheless, consideration of alternative futures underlines the great uncertainty that confronts decision-makers.

The negative effects or symptoms of uncertainty often include delay, hesitation and vacillation. These should not be confused with active strategies of delaying or

103 Freedman, *Revolution*, p. 68.
104 Gray, *Bloody*, p. 159.
105 *On War*, I, 3, p. 131.
106 *On War*, I, 3, p. 131.
107 *On War*, I, 3, p. 125.
108 Carr, *What is History?*, p. 97.

avoiding battle, such as the Periclean or Fabian approach.[109] Moreover, for those not inured to uncertainty, the psychological consequences of uncertainty can so relentlessly sap confidence as to induce great anxiety, if not outright behavioural paralysis – consider for instance Ludendorff's nervous breakdown in 1918. Clausewitz notes how 'most generals, when they ought to act, are paralysed by unnecessary doubts'.[110]

Irresolution and reluctance to act, Clausewitz remarks, particularly harm those on the offensive, which generally requires continuous momentum, quick decision and exploitation of the initiative: if the commander demands absolute certainty he will inevitably delay until the optimum moment for action has passed. A mind racked by uncertainty will always want that little bit more time to weigh up alternatives and check all available information before acting. Much to his consternation, in 1809 Clausewitz had witnessed the consequences of such vacillation when Napoleon was close to collapse against the Austrians at Aspern-Essling in 1809 and the Prussian King, Frederick William failed to act: his hopeless message to the Austrians was, 'One more victory and I am with you'.[111] The instances of such fatal indecision are legion. A notable example is that of Union commander George McClellan during the American Civil War, who refused to attack even when armed with the enemy's campaign plan and with circumstances favouring action.

Equally, however, uncertainty can lead to overconfidence, even in the face of an enemy's massive numerical superiority because, after all, fortune purportedly favours the bold. Clausewitz suggests that a balanced reaction to uncertainty is required: 'While one man may lose his best chance through timidity … another will plunge in head first and end up looking as dazed and surprised as if he had just been fished out of the water'.[112] As Clausewitz recognised, it is part of human nature to place confidence in one's own success. It was this sentiment that lay behind Clausewitz's thinking in his 'Political Declaration' of 1812 in which he argued that even against the odds, Prussia could find courage in despair: 'It is true that the *probability* of success is against us … How can anyone demand the probability of success! It is enough that success is not *impossible*; whoever asks for more contradicts himself'.[113]

The overriding consequence of this chronic uncertainty is that almost all major decisions will be largely based on 'probability and inference' and shaped by the psychological disposition of and nature of the relationships between the commanders and statesmen at the time rather than any universal standard. Clausewitz states that, 'Circumstances vary so enormously in war, and are so

---

109 Clausewitz notes that, 'All campaigns that are known for their so-called temporising, like that of the famous Fabius Cunctator, were calculated primarily to destroy the enemy by making him exhaust himself'. *On War*, VI, 8, p. 460.

110 *On War*, III, 1, p. 209.

111 Parkinson, *Clausewitz*, p. 118.

112 *On War*, VII, 22, p. 689.

113 Clausewitz quoted in Paret, *Clausewitz*, p. 302.

indefinable, that a vast array of factors has to be appreciated – mostly in the light of probabilities alone'.[114] As such, he is adamant that reliable prediction is impossible: 'absolute, so-called mathematical factors never find a firm basis in military calculations'.[115] In a more objective sense, it is this unpredictability which entails that, as he puts it, 'no prescriptive formulation universal enough to deserve that the name of law can be applied to the constant change and diversity of the phenomena of war'.[116]

Incidental friction

If the Enlightenment military thinkers had conceived of military institutions as 'mathematical reason in action' or as great machines 'where each part fulfilled its prescribed function, with no waste motion and no supernumerary cogs',[117] then it was Clausewitz who shrewdly introduced the logical, adverse corollary of the metaphor through his concept of friction. Friction was notably absent from the military machines envisaged by the military *philosophes*, as if they operated in a vacuum. Clausewitz endeavoured to correct this oversight. In *The Principles of War*, Clausewitz compares the conduct of war to 'the workings of an intricate machine with tremendous friction'.[118] In *On War*, he describes how, 'Iron will power can overcome this friction … but of course it wears down the machine as well'.[119] Yet, as he makes clear in his study of *The Campaign of 1812*, the analogy does not entirely suffice because, unlike in mechanics, the friction 'cannot be reduced to a few points, but is everywhere in contact with chance'.[120]

As Barry Watts has persuasively shown, the concept of friction can be understood in both its 'incidental' and 'general' manifestations. In the first case, friction needs to be understood as a separate phenomenon explaining a particular pervasive feature of war. The concept appears fairly early in Clausewitz's thinking and rapidly became a major theme in *On War*.[121] Incidental friction is for Clausewitz an 'unseen, all pervading element' and a 'force that theory can never quite define', but its general features are quite clear, particularly when accompanied by the powerful metaphors he provides.[122] It is a factor that reduces the efficiency of the whole and 'impedes activity' through 'natural inertia' as 'countless minor incidents

114 *On War*, I, 3, p. 130.

115 *On War*, I, 1, § 21, p. 97.

116 *On War*, II, 4, p. 176.

117 Henry Lloyd had often described the army as a great machine. Gat, *Military Thought*, p. 73.

118 Clausewitz, *Principles*, p. 61.

119 *On War*, I, 7, p. 138.

120 Clausewitz, 'Campaign of 1812', p. 165.

121 Watts, *Clausewitzian Friction*, pp. 7–8.

122 *On War*, I, 7, p. 140.

... combine to lower the general level of performance'.[123] A defining feature of the concept is its *cumulative* nature, in the sense that a multitude of small 'difficulties' leads to much larger problems. It conveys, as Beyerchen notes, the 'amplification of a microcause to a macro-consequence, in a kind of cascade of things gone wrong'.[124] Clausewitz states that 'difficulties accumulate and end by producing a kind of friction that is inconceivable unless one has experienced war'.[125]

This cumulative effect results largely from the nature of militaries, which like complex and intricate machines, are made up of many individual parts each with the potential to produce friction and make, as Clausewitz puts it, 'the apparently easy so difficult'.[126] It is these factors that make war 'like movement in a resistant element' and here Clausewitz employs the fitting analogy of walking in water: 'Just as the simplest and most natural of movements, walking, cannot easily be performed in water, so in war it is difficult for normal efforts to achieve even moderate results'. Friction is composed of all manner of difficulties, whether they be fatigue, adverse weather, insufficient provisions and so forth: he states that it would take volumes to cover them all. Moreover, Clausewitz explains that friction is 'everywhere in contact with chance, and brings about effects that cannot be measured, just because they are largely due to chance'.[127] He also notes that, 'Often we encounter obstacles which were impossible to foresee'[128] and that you can never know when one of the parts of the whole will 'chance to delay things or somehow make them go wrong'.[129] Friction, like chance, is not something that can be prevented or forestalled – one can only attempt to limit its effects and be mindful of its universal presence.

General friction

So far we have discussed the prominent concepts in theoretical isolation to arrive at a firmer understanding of their distinctive features. However, Clausewitz also believed that it was possible for these factors to be conceived of in a unitary sense. All these factors, along with danger – the 'debilitating element' – and physical exertion, correspond with respect to their 'restrictive effect' in war and as Clausewitz argued, for this reason 'can be grouped into a single concept of *general friction*'.[130] There are clearly many overlaps in relation to the individual concepts discussed above, such as chance and uncertainty, but the idea of general friction can be regarded as a means to collectively perceive the various ways in which

123 *On War*, I, 8, p. 141, VIII, 2, p. 702 and I, 7, p. 138.
124 Beyerchen, 'Nonlinearity'.
125 *On War*, I, 7, p. 138.
126 *On War*, I, 7, p. 140.
127 *On War*, I, 7, p. 139.
128 Clausewitz, *Principles*, p. 64.
129 *On War*, I, 7, p. 139.
130 *On War*, I, 8, p. 141. Emphasis added.

activity in war is impeded everywhere and at every stage. The basic nature of the force being described is essentially analogous to that associated with incidental friction, but here it encompasses the whole range of factors associated with action in war.

In Watt's excellent study, he draws attention to other potential sources of general friction, which are identifiable in Clausewitz's wider work. In particular, he draws attention to the political constraints on the use of force and the intractable difficulty of matching military means to political ends.[131] Whatever factors we include, the overarching thrust is clear, as is the extent to which this friction pervades the entire gamut of war, from those determining the political object to individual soldiers. While identification of this intrinsically restrictive force in war may seem to give cause for outright negativity, this would be a mistaken conclusion. Given war's interactive nature, it is always necessarily relative. This tremendous friction in war is something best offset through the efforts of the energetic and creative genius – nature's compensation for uncertainty.

Playing the lethal game of war

War, states Clausewitz, can be compared to a game of chance and thus 'guesswork and luck come to play a large part in war'.[132] In its danger and seriousness war is perhaps most akin to Russian Roulette, but this does not convey the extraordinary talents required, which are perhaps more accurately captured in the analogy of poker, where it is 'a matter of skill as well as odds'.[133] In the latter, one's fate, while still subject to the laws of probability, is more heavily dependent on choice and accurately reading one's opponents, while in the former one's fate is left purely to chance. In his analysis of Napoleon's 1799 campaign, Clausewitz states that, since war 'always has something of the nature of a game, the conduct of war cannot avoid this element at every stage, and the commander, who has little inclination for the game, will, without anticipating it, be left behind the line and will fall into deep debt in the great account book of military success'.[134]

Yet, what makes war such a unique game of chance is the mixture of uncertainty with ever-present danger, fear, physical exertion and, for the commander and policymaker, an enormous burden of responsibility: so 'luck in war is of a higher quality than luck in gambling'.[135] Far from being a mere pastime, Clausewitz asserts, war is not a place for 'irresponsible enthusiasts. It is a serious means to a

131 Watts, *Clausewitzian Friction*, pp. 30–32.

132 *On War*, I, 1, § 20, p. 96.

133 Alan Beyerchen, 'Clausewitz and the Non-Linear System of War: Systems of Organized Complexity', in Strachan and Rothe, *Clausewitz*, p. 53.

134 Quoted in Hew Strachan, *Carl von Clausewitz's On War: A Biography* (New York: Atlantic Monthly Press, 2007), p. 144.

135 *On War*, II, 5, p. 195.

serious end'.[136] *On War* is essentially geared towards preparing commanders for this 'relentless struggle with the unforeseen'.[137] Yet, he recognised that no book can provide the requisite abilities needed to face such an ephemeral foe. As Paret explains, Clausewitz believed it is 'the creative employment of intellectual and psychological strengths that alone can overcome friction, exploit chance, and turn the imponderables into an asset'.[138] The qualities of genius represent the counterpoise to the effects of chance and uncertainty.

Here a definite dualism is apparent concerning the inherent interrelationship between the condition of uncertainty on one hand and the consequent human reaction on the other: they form two sides of the same coin. The relationship can often appear somewhat unbalanced, with chance dictating terms to submissive human will. However, a more accurate conception is one of mutual dependence, even symbiosis, in so far as human creativity, free will, imagination and choice are fundamental causes of the condition in the first instance. Put simply, the fact that the enemy is a living and thinking entity with a will of its own is a major source of chance and uncertainty – war is never activity aimed at inanimate matter.[139] As Gray states, 'War is a duel and the enemy may prove uncooperative'.[140] Even absent the unpredictability of the external physical sources of chance, humans appear to be both problem and solution, especially when war is viewed properly, as Clausewitz insisted it should, as an interactive, multilateral whole.

Genius: the creative talent

As noted earlier, in his description of genius, Clausewitz was echoing many of those attributes which Machiavelli had identified in his concept of *vitrù*, while examining them in a more systematic fashion. Considering the fact that the science of psychology was only in its infancy, as Clausewitz acknowledged, his penetrating analysis in Book 1, Chapter 3 is a jewel of his work. Clausewitz prefaced his discussion with the observation that genius is the combination of certain qualities of mind and temperament that are specific to war and that may not necessarily be appropriate elsewhere, emphasising that the 'gifts' of military genius are those which pertain to the unique climate of war. The qualities required are different from those of pure understanding, intellect or even technical knowledge. Rather, for Clausewitz, genius is a complex mix of intellect and emotional qualities, 'a blend of brains and temperament'.[141]

Genius in war, he explained, must be grounded in courage. However, if the old adage confirms that discretion is the larger part of valour, then Clausewitz

136 *On War*, I, 1, § 23, p. 98.
137 *On War*, I, 3, p. 117.
138 Paret, *Clausewitz*, p. 373.
139 *On War*, II, 3, p. 173.
140 Gray, *Bloody*, p. 46.
141 *On War*, I, 3, pp. 130–31.

emphasised a form of courage that was not wild, but controlled and educated. Courage encompasses such traits as boldness, presence of mind (such as dealing with the unexpected or keeping one's nerve), strength of will (in the face of anxiety), as well as the ability to take risks and to trust in fortune. Purely rational thinking would struggle to cope with or be overwhelmed by the lack of evidence and certainty, so a form of intellectual instinct or *coup d'oeil* – 'the quick recognition of a truth' – reinforced by determination was necessary, which would enable the commander to effectively turn knowledge into ability.[142] The military genius would also need to possess adaptability and creativity, combined with 'great strength of character, as well as lucidity and firmness of mind … in order to follow through steadily, to carry out the plan and not to be thrown off course by a thousand diversions'.[143] All this requires great energy or what Clausewitz terms a spirit of endeavour.[144] Importantly, and recalling Machiavelli, genius is not just about coping with uncertainty but actively exploiting it: one can attempt to turn uncertainty to one's own advantage.

In the face of the growing complexity of war, a notable feature of modern warfare has been the increasing collegiality of command and the rise of sprawling military bureaucracies and planning staffs. During the nineteenth century, general staffs emerged which increasingly took responsibility for many of the decisions that would formerly have taken place in the commander's mind.[145] As a result, genius has become increasingly dissociated from the individual and supplanted by considerations of bureaucratic efficiency and consensual decision-making. These changes have primarily been functional imperatives in response to the expanded dimensions and varieties of knowledge required in modern warfare. However, these developments have encouraged practices and behaviour often unsuited to dealing with the uncertainties of war; not least excessively rigid planning.

For instance, in the decades prior to 1914, planning became an obsession of European staffs, yet, as events were to reveal, brilliant execution of the wrong plan can be extremely dangerous. The French strict adherence to 'Plan XVII' led to dispositions and movements almost wilfully unsuited to unfolding events.[146] This is not to argue that detailed planning is unimportant. Rather, the danger is that rigid plans can become confused with future reality, when in fact no blueprint, however sophisticated, can entirely account for the unpredictable events that will inevitably be encountered. Dogmatic planning can encourage a mistaken belief that victory can be orchestrated in advance. Gray quotes Admiral Wylie who stated that 'planning for certitude is the greatest of all military mistakes'.[147] In this

142 *On War*, I, 3, pp. 130–31 and II, 2, pp. 170–71.

143 *On War*, III, 1, p. 209.

144 *On War*, I, 3, pp. 121–2.

145 General staffs were increasingly recognised as the 'brain of the army'. Bond, *Pursuit*, p. 79.

146 See Gat, *Military Thought*, pp. 382–440.

147 Quoted in Gray, *Bloody*, p. 286.

regard, the genius of the overall commander can be the crucial factor preserving the element of *coup d'oeil*, flexibility and adaptability in the face of failed plans and unexpected occurrences.

Military virtues

So far we have concentrated principally on the higher command in war, but we must also consider relevant features evident at lower levels. Indeed, one of the central problems in war is how to produce soldiers willing to risk their lives in the 'debilitating element' of battle and stand-fast in the face of potential death.[148] As Dyer notes, 'Soldiers in battle, however stable they may appear, are always a potential mob capable of panic and flight'.[149] Desertion, mutiny and the collapse of morale are the nightmares of generals. Consequently, throughout history, armed groups have developed a diverse range of responses (consciously and unconsciously) to overcome such problems, even if they sometimes solidified into cultural habits that no longer truly served their purpose.[150] Nevertheless, where the latter occurred, military necessity, and the requirement to adapt or die, generally engendered the discarding of out-dated forms.

Clausewitz discussed such issues through the concepts of 'military virtues' and 'military spirit', which in fighting forces he regarded as the 'steering quality, [the] refinement of base ore into precious metal'.[151] Today, many of the qualities which Clausewitz recognised as vital in command are increasingly required by soldiers at all levels as closed order formations have been replaced by more open fighting demanding the 'free play of intelligence, the clever merging of boldness with caution'.[152] Clausewitz believed that foremost amongst the virtues that can reduce friction in fighting forces is simply combat experience or 'long familiarity with danger', and described it as 'the only lubricant that will reduce the abrasion of friction.[153] He noted how, 'Habit hardens the body for great exertions, strengthens the heart in great peril, and fortifies judgement against first impressions'. For militaries without direct experience of war, efforts are often made to provide troops with training approaching true battlefield conditions, although, as Clausewitz recognised, 'peacetime maneuvers are a feeble substitute for the real thing'.[154]

The development of detailed military doctrine has become a central means to reduce uncertainty given the complexity and technicality of modern warfare. Dyer notes how militaries are often criticised for reducing action to routines and rules, but 'all it amounts to in practice, is a desperate and partially successful

148 *On War*, I, 4, p. 133.

149 Gwynne Dyer, *War: The Lethal Custom* (New York: Carroll & Graf, 2005), p. 21.

150 See Martin van Creveld, *The Culture of War* (New York: Ballantine Books, 2008).

151 *On War*, III, 5, p. 221.

152 Parkinson, *Clausewitz*, p. 127.

153 *On War*, I, 4, p. 133 and I, 8, p. 141.

154 *On War*, I, 8, p. 141.

attempt to reduce the immense number of variables'.[155] This is true, but doctrine can also be greatly counter-productive due to the rigidity of thinking it encourages. Clausewitz warned that, in war 'theoretical *directives* tend to be less useful here than in any other sphere'.[156] Of course, in 1806 Clausewitz had witnessed the 'most downright stupidity to which methodicism ever led'.[157] It is little wonder that the new Prussian Army Regulations of 1812, which Clausewitz helped draft, emphasised flexibility, initiative and exploitation of the unexpected.

Discipline and drill represent historic practices that derive from the imperative to 'stand fast in the face of the enemy ... without giving way to the natural impulses of fear and panic'.[158] Clausewitz notes how such practices form an 'army that maintains its cohesion under the most murderous fire; that cannot be shaken by imaginary fears and resists well-founded ones with all its might'.[159] Strict discipline has often been employed where soldiers have been reluctant to fight. At times, outright coercion has been ruthlessly applied. Indeed, the term 'decimation' derives from the Roman punishment of executing every tenth man from units who were defeated or humiliated in battle.[160] Countless other ways of preparing troops for the uncertainties of war have been adopted. Formalised hierarchies ensure orders are followed. Insignia, flags (or 'colours'), ceremonies and histories serve to foster a culture of duty, valour and group belonging necessary to maintain cohesion. Clausewitz describes how fighters often think of themselves as a member of a guild'.[161] Also, at a more personal level, if, as Clausewitz stated, 'what is most needed in the lower ranks is courage and self-confidence',[162] then soldiers throughout history have adopted diverse ways of reconciling themselves to the danger of war, such as taking drugs, drinking alcohol, saying prayers, covering themselves in war paint and even finding consolation in Stoic philosophy.[163]

These examples are intended to emphasise the extent to which many of the iconic images and practices we associate with war derive largely from its uncertain nature. But, as Clausewitz believed, none of these measures can truly prepare the soldier for the sheer emotional intensity and confusion of *actual* combat. As he states, 'discipline, skill, goodwill, a certain pride, and high morale ... have no strength of their own. They stand or fall together. One crack and the whole thing

---

155 Dyer, *Lethal Custom*, p. 12.

156 *On War*, II, 2, p. 161.

157 Parkinson, *Clausewitz*, pp. 311–12.

158 Geoffrey Parker, 'The Western Way of War', in Geoffrey Parker (ed.), *The Cambridge History of Warfare* (New York: Cambridge University Press, 2005), p. 3.

159 *On War*, III, 5, p. 220.

160 Murray et al., *Making of Strategy*, pp. 60–61.

161 *On War*, III, 5, p. 219.

162 *On War*, II, 2, p. 162.

163 See Sherman, *Stoic Warriors*, pp. 1–9.

goes, like glass too quickly cooled'.[164] Like the best lubricants, they can only reduce the friction of war, but not entirely eradicate it.

Implications

The externality of chance and uncertainty?

Bassford argues this element of the trinity is the only one which is exclusively concerned with the 'objective externality' of war, which for him consists of, 'The physical world (… in short, everything we cannot alter at once by merely wishing) *and* of the personalities, capabilities, hopes [etc] … the human ecology within which the participants' perceptions, plans, and actions must co-evolve'.[165] This conception is useful and captures a vital dimension of this element. It forces us to conceive of 'chance and probability' as something 'out there', as characterising the conditions within which action takes place in war.

Yet, this perspective, important and correct to a point, does not suffice. It ignores its vital complementary aspect which, as we have seen, is in fact an internal feature of the actors involved in war. We have considered the extent to which concepts such as chance and uncertainty are not only a function of an external reality, but also exist as a characteristic of actors' internal predicament, intellectual capacity or psychological condition: it is not so much that the objective environment is uncertain, but rather the fact that actors in war experience the inner feeling of uncertainty that is important. To reduce this tendency to an examination purely of war's external reality would therefore be somewhat mistaken. Nevertheless, an external 'concrete reality' is certainly implicated in the meaning. Such factors as the weather or the actions of the enemy are, of course, external manifestations of chance.

In an imaginary world of God-like omniscient beings the word chance would presumably not exist, because all effects would have a definite cause and the outcome of any action would be known in advance – in such a world, perhaps Creveld is right to suggest that there would be no war (if the result is known in advance there would be no point in fighting).[166] Yet, war *is* a human activity, with all the intellectual shortcomings, passions and unpredictable behaviour that entails: some events and outcomes will always be unexpected. While Bassford is perhaps correct to present chance as an external element of war, because that is how it *appears*, chance is also partly a function of human nature, describing something that emanates from inside ourselves and is not simply 'out there'.

---

164 *On War*, III, 3, p. 222.

165 Bassford, 'The Primacy of Policy', p. 89.

166 Creveld, *Transformation*, p. 171.

Ambiguity

The second central implication to outline is the inherent ambiguity of this tendency. Enlightenment theorists prior to Clausewitz viewed chance in an almost universally negative light and one's instinct might be to support such a view. However, Clausewitz developed a more realistic and balanced conception of these forces, both in terms of how they are experienced by humans and in terms of their objective impact on war. Regarding the latter, Echevarria stresses that friction and chance are not in themselves forces which do or do not lead to extremes; rather they are the medium in which such interactions develop. Friction cannot itself stop war from escalating – 'at times the forces that drive war are strong enough to overcome inertia or friction; at other times they are not'.[167] While it is one of the forces that makes absolute war nothing but a logical fantasy, in war in reality it is utterly ambiguous in relation to escalation.

From a subjective strategic perspective, and comparable to Machiavelli's fortune, this tendency is neither a strictly positive or negative force. Chance, uncertainty and friction will invariably be experienced by all sides in war. Herbig notes that, 'chance remains neutral' and it is best perceived as ambivalent in terms of its impact on belligerents. Chance events can create possibilities and opportunities to be exploited or equally dash the best laid plans. It can cause the complete failure of a campaign, such as the 'divine wind' which twice frustrated the thirteenth-century Mongol invasion of Japan,[168] or decisively support a surprise move. Given a rough equality between belligerents in relation to these factors, this is where military genius, political acumen, effective intelligence techniques, combat experience, superior morale, flexible planning, rigorous training and other factors can play their part in taming or exploiting the uncertain climate of war.

Friction may never be eliminated entirely and, objectively, it may be considered a negative or restrictive force, but when considered relative to the enemy, if you can reduce the friction of your own side to a greater extent, significant advantages can accrue. Uncertainty may paralyse one commander, but it may appear to another as a world full of possibilities within which creative genius can take wing. The same idea was expressed by Thucydides when he stated, 'That imponderable element of the future is the thing which counts in the long run, and, just as we are most frequently deceived by it, so too it can be of the greatest possible use to us'.[169] Indeed, Clausewitz himself could not be clear enough on this subject: 'we should not habitually prefer the course that involves the least certainty … There are times when the utmost daring is the height of wisdom'.[170] Again, this echoes Machiavelli who included a chapter in his *Discourses* entitled, 'Results are often

167 Echevarria, *Clausewitz*, p. 117.

168 Harold A. Winters, *Battling the Elements: Weather and Terrain in the Conduct of War* (Baltimore: John Hopkins University Press, 1998), pp. 9–15.

169 Thucydides, *Peloponnesian War*, p. 301.

170 *On War*, II, 5, p. 195.

obtained by Impetuosity and Daring which could never have been obtained by Ordinary Methods'.[171]

Chance and practical, means-oriented activity

Thirdly, although perhaps obvious, it is important to underline that this tendency relates largely to the conditions in which practical, goal directed activity in war is conducted, whereby means are applied through creative human activity. While it may describe external and internal realities (such as, respectively, surprise enemy moves or misleading ethnocentric assumptions), it does so in relation to what such things entail for the behaviour and potential of the active participant in war. Here we are referring to such activity in the very broadest sense: political and diplomatic initiatives, strategic decision-making, actual combat and so on, incorporating activity from the tactical to political-strategic levels.

Strategy itself is at heart a practical activity that seeks to direct resources towards the attainment of certain goals. In this respect, it is a dynamic activity that takes place in the conditions of uncertainty described above and requires great talent. Of course, strategy relates to and must take account of all aspects of the trinity, but here we want to stress the extent to which it will be conditioned by the inescapable factors associated with this tendency. It is not surprising that two modern authors could define strategy as 'a process, a constant adaptation to shifting conditions and circumstances in a world where chance, uncertainty, and ambiguity dominate'.[172]

*

Theorists prior to and since Clausewitz have approached the concept of chance in a manner which tends to be either parochial, melodramatic or smugly dismissive. Following Machiavelli, Clausewitz developed a more realistic conception. Moreover, he was impressed by the demands chance and uncertainty make on those faced with such a reality: what we might call the 'psychological consequences of chronic uncertainty'.[173] His fascination with the creative talents such conditions call forth is revealed when he states that, 'Although our intellect always longs for clarity and certainty, our nature often finds uncertainty fascinating. It prefers to daydream in the realms of chance and luck rather than accompany the intellect on its narrow and tortuous path of philosophical inquiry and logical deduction'.[174] War represents the ultimate testing ground of the human creative spirit. Individual soldiers faced with the fear of death or commanders holding the fate of nations in

171 Machiavelli, *Discourses*, p. 519.

172 Williamson Murray and Mark Grimsley, 'Introduction: On Strategy', in Murray et al., *The Making of Strategy*, p. 1.

173 Herbig, 'Chance and Uncertainty', p. 107.

174 *On War*, I, 1, § 22, p. 97.

their hands must make decisions in the midst of relentless uncertainty and crushing responsibility. What greater test can the human will face?

Chance and uncertainty should not be conceived as being everywhere, all of the time – there are observable linear cause and effect relationships in war. However, uncertainty is central to the nature of the phenomenon, along with the measures humans employ to overcome it. Modern developments, such as those associated with high-tech systems or professional intelligence agencies have not rendered these insights obsolete. If anything, these ideas may be even more relevant to modern warfare which takes place in multiple dimensions, employs a mind-boggling mix of old and new technologies, is pervaded by endlessly complex political, legal and ethical imperatives – refracted through an ever-present mass media – and involves multiple actors, diverse groups and complex bureaucracies.

Chapter 6
Passion: The Blind Natural Force

This tendency is described as being composed of 'primordial violence, hatred and enmity, which are to be regarded as a blind natural force'.[1] Clausewitz then refers to this element as 'passion'. Centuries earlier, Thucydides had described in a similar fashion the sort of deep-seated and atavistic forces suggested by Clausewitz: 'with the ordinary conventions of civilised life thrown into confusion, human nature, always ready to offend even where the law exists, showed itself proudly in its true colours, as something incapable of controlling passion.'[2] Humans, being merely animals who have developed the capacity for reason, always retain the potential to display the characteristics deeply embedded in their nature, while that nature is essentially ambiguous in its manifestation, neither intrinsically good nor bad: violent passion is just one of its potential forms. We sometimes need to be reminded just how fragile, in reality, the power of reason is when fear predominates, when basic emotions are aroused and when perceived obstacles to personal gain are removed.

These primordial aspects of human nature are conceived as things which neither science nor reason can fully control. If we imagine a world without such traits evident, or at least potential, in the human species, then we enter a world of make-believe inhabited by denying anthropomorphobics.[3] To postulate a purely rational world inhabited by automatons following the mechanical dictates of reason would fail to capture the reality of war; but, moreover, war is precisely the sort of activity in which such instinctive characteristics have the potential to usurp human behaviour. As Creveld notes, 'Nothing is more likely than the terror of war to cause rationality to go by the board, nor is anything more conducive to make even the most even-minded start behaving somewhat strangely'.[4] Likewise, Clausewitz remarks how fighting 'will stir up hostile feelings … That is only human (or animal, if you like), but it is a fact'.[5] To acknowledge the play of such forces was crucial for Clausewitz and warranted inclusion as a clearly defined element of war.

---

1 Carl von Clausewitz, *On War*, trans. and ed. Michael Howard and Peter Paret (New York: Everyman's Library, 1993), I, 1, § 28, p. 101. Emphasis added.

2 Thucydides, *History of the Peloponnesian War*, trans. Rex Warner (London: Penguin, 1972), p. 245.

3 Anthropomorphobia refers to those who deny that humans share any common traits with animals.

4 Martin van Creveld, *The Transformation of War* (New York: Free Press, 1991), p. 89.

5 *On War*, II, 2, p. 159.

So, what compelled him to incorporate such ideas into his theory, what did he really mean when he spoke of 'primordial violence', does contemporary scholarship support these claims and how are we to understand the operation of this tendency in reality? The answers to these questions are not obvious and grasping Clausewitz's meaning requires detailed consideration. This tendency is the one most often neatly side-stepped by commentators, either perhaps because they feel its implications are too obvious or the concept is seen as too amorphous to capture in scholarly theoretical work. We must attempt to attain a firmer understanding.

Influences

We will be better placed to present an accurate description of the meaning of this tendency if we are aware of some of the influences that contributed to the development of Clausewitz's thought. This tendency is not neatly and conveniently associated with specific aspects of his intellectual milieu. Rather, we will depict the broad context within which Clausewitz's ideas germinated, developed and coalesced and yet will not hesitate to emphasise strong connections between a concept and its roots where apparent.

Intellectual influences

We noted in Chapter 2 that Clausewitz's methodology was heavily indebted to the Enlightenment and the broad standards of rational enquiry it promoted. Yet, where the steady ripples of reason lapped up against the shores of *On War*, so the solid ground of the work truly began to rise out of these calm waters, giving way to a rugged and powerful landscape more in the spirit of the poetry of Wordsworth and Coleridge, than of rococo formality or neo-classical congruity. In this respect, Clausewitz represents an original page in the history of military thought.

By the end of the eighteenth century an oppositional intellectual movement to the Enlightenment had emerged, which, in broad terms, was critical of the world-view associated with the 'mechanistic rationality'[6] of the Enlightenment. This so-called 'Counter-Enlightenment' – or 'the culture of feeling'[7] – would be as, if not more, diverse than the movement it was reacting against. Its origins have often been traced back to the thought of Jean-Jacques Rousseau who resented the marginalisation of 'feeling' due to the 'cold calculation' of Enlightenment reason.[8] Rousseau arguably sowed the seeds of the cultural backlash that would find its

6 Michael Howard, *The Invention of Peace and the Reinvention of War* (London: Profile Books, 2001), p. 112.

7 Tim Blanning, *The Pursuit of Glory: Europe 1648–1815* (London: Penguin, 2007), p. 513.

8 Eric Hobsbawm, *The Age of Revolution: Europe 1789–1848* (St Ives: Abacus, 1999), p. 300.

most striking expression in the output of the German Movement. Its leading lights were Clausewitz's contemporaries, many of whom he met, corresponded with or at least read: their influence on his thought was significant.

Immanuel Kant – whom we know Clausewitz was familiar with through his attendance of the lectures of the Kantian populariser, Johann Kiesewetter, while at the Berlin War Academy[9] – represented a crucial link between the Enlightenment and the German Movement. Following Kant's refusal to accept the objective existence of the 'thing-in-itself', a philosophical tradition emerged in Germany which promoted the subjective self and its mental faculties.[10] The element of anti-rationalism in Kant's 'critical project' was partly reflected in the influential literature of the *Sturm und Drang*[11] movement in the closing decades of the eighteenth century, whose leading proponents were Hamann, Klinger, Möser, Lavater and the early Goethe and Schiller. As Tim Blanning notes, 'this was very much an angry young man movement against what was perceived as the stultifying rationalism and classicism of the older generation'.[12]

Their writings expressed the rich, vital and dynamic nature of human reality through the will of charismatic central personalities such as Goethe's *Werther*[13] or the characters in Schiller's dramas, *The Robbers* and *Wallenstein*. In their works they extolled extremes of emotion and tormented subjectivity. The freedom, will, consciousness and wholeness of self – 'the boundless potential of the human spirit'[14] – was revealed through imaginative insights into man as a creative, unique and feeling being. Decrying the mechanistic unfeeling formality of rationalist behaviour, Goethe's Werther cries, 'Ah, you sensible people! … Passions! Intoxications! Insanity! You are so calm and collected, so indifferent, you respectable people … '[15] Clausewitz, as with much of German society, was swept up in 'Werther-mania'.[16] Goethe's *Faust* expresses the desire of the frustrated

9 Peter Paret, *Clausewitz and the State: The Man, His Theories, and His Times* (Princeton: Princeton University Press, 1985), p. 71.

10 Roger Scruton, *A Short History of Modern Philosophy* (London: Routledge, 2005), p. 164.

11 The movement took its name from a play by Freidrich Maximillian von Klinger.

12 Blanning, *Pursuit of Glory*, p. 518.

13 Goethe's novel, *The Sufferings of Young Werther*, was the first international best-seller written by a German. Blanning, *Pursuit of Glory*, p. 519.

14 F. J. Lamport, 'Introduction' in Friedrich Schiller, *The Robbers and Wallenstein* (London: Penguin, 1979), p. 7.

15 Johann Wolfgang von Goethe, *The Sorrows of Young Werther* (London: Penguin, 1989), p. 61.

16 *Werther* was one of the subjects discussed on the occasion of his first meeting with his future wife Marie. Azar Gat, *A History of Military Thought: From the Enlightenment to the Cold War* (Oxford: Oxford University Press, 2001), p. 184.

intellectual to 'leave behind the cerebral world of mere thought and embrace with all his senses the full human lot' by making a pact with the devil.[17]

The radical message contained in these works is unmistakable in Clausewitz's trinity. Moving away from the largely 'dead abstractions'[18] of Enlightenment military thought, Clausewitz strove to place living man firmly at the centre of his theory, which would reveal the fundamental subjective forces that permeate war. He notes how, 'Theorists are apt to look on fighting in the abstract as a trial of strength without emotion ever entering into it', when in fact, 'All these and similar effects in the mind and spirit have been proved by experience: they recur constantly, and are therefore entitled to receive their due as objective factors'.[19] The authors of the German Movement focused on the real human impulses that drive men to action – be it superstition, jealousy, ambition, despair, greed or revenge – rather than an appeal to more noble or strictly rational sentiments. In a similar vein, Clausewitz did not shy away from acknowledging these base instincts and the extent to which they can dominate human behaviour; not least the ever-present potential for evil which, as Schiller puts it, rejects 'the iron yoke of mechanical laws,'[20] hence necessitating the constant vigilance of would-be victims.

Such themes were developed and taken to their extremes by the German Romantics, many of whom moved in the same social circles as Clausewitz.[21] The final rejection of neo-classicism, the stress on feelings and the elevation of the spiritually free, creative individual were some of the dominant features of the Romantics, such as the Schlegel brothers and Novalis. Given the pervasive and deep impact of these writers and Clausewitz's intimate association with them, it is accurate to assume that the many sections of Clausewitz's work that emphasise emotional forces owe a debt to the Romantic outlook. Yet, Paret is correct when he notes how Clausewitz did not follow the Romantics in their surrender to emotion[22] and Gat states that it is clear that he rejected both their 'idealism and mysticism'.[23]

It should also be remembered that not all those we might broadly designate as 'Romantic' subscribed to the extremes of the movement or at least moved away from them in later life, such as was the case with the later Goethe and Schiller. Often such writers were more concerned with the interplay of reason and emotion, rather than with any zero-sum espousal of one over the other. Schiller is of particular interest. In his more mature works he addressed 'the malaise of modern

17 Nicholas Boyle, *German Literature: A Very Short Introduction* (Oxford: Oxford University Press, 2008), p. 52.

18 Gat, *Military Thought*, p. 180.

19 *On War*, II, 2, p. 158.

20 Schiller, *The Robbers*, p. 56.

21 Roger Parkinson, *Clausewitz: A Biography* (New York: Stein and Day, 1979), pp. 90–91. Clausewitz met August Wilhelm Schlegel when were both were guests of Madame de Staël in Coppet, Switzerland in 1807.

22 Paret, *Clausewitz*, p. 149.

23 Gat, *Military Thought*, p. 185.

man, analysing him into his rational and sensuous components' concluding that 'the seeds of rational perception will wither where they fall unless the soil has been prepared by the emotions and imagination'.[24] As exemplified in the trinity, it is the complex, dynamic interplay between reason and emotion that fascinated Clausewitz so much, rather than any strict dichotomy between the two.

Indeed, returning to Clausewitz's debt to the Enlightenment, perhaps somewhat counter-intuitively we can identify its influence here also. It is, of course, a gross simplification to characterise the Enlightenment as abandoning all sensibilities in favour of a cold rationalism. In fact many of its leading lights, particularly its earlier representatives, were greatly concerned with the place of emotions in their philosophical systems. Moreover, Clausewitz probably owes more to the early Enlightenment than its later more mechanistic, positivistic manifestations; perhaps more than to the wild Romantics mentioned above. In one of his notes written around 1818, Clausewitz acknowledged his debt to Montesquieu, primarily in relation to the style of his presentation. Yet, a deeper affiliation may be posited: as Paret notes, Clausewitz 'must have valued Montesquieu's skepticism, specificity, and recognition of the importance of irrational factors'.[25]

Men such as Adam Smith devoted a great deal of their writings to the study of emotion and sentiment.[26] Smith's famous invisible hand did not operate purely as a result of individual rational calculations, but more due to behaviour governed by instincts and emotions; as John Gray has noted, Smith was 'suspicious of the intellect when it operated without regard for sentiment'.[27] Such views are inconsistent with classical political economy, just as Clausewitz's conception of war does not correspond with his excessively rationalistic antecedents. Outcomes in war cannot simply be explained by the application of purely rational methods, but must account for the play of emotional, cultural and individual psychological factors.

We should perhaps not think of Clausewitz's conception of this tendency as corresponding to the Romantic notion of a strict dichotomy between emotion and reason, whereby the two are fundamentally at odds with one another, but rather as a constant interplay between the two. This insight has three implications worth noting. First, simply because Clausewitz emphasised such emotional forces does not necessarily imply a complete subscription to the Romantic outlook – early enlightenment thinkers were perfectly comfortable with such ideas. Second, the existence of passions in his theory does not necessarily imply a rigid incompatibility with those aspects concerned with the use of reason; the two are not mutually exclusive. Third, perhaps somewhat paradoxically, emotion is something which a

---

24 Schiller, *The Robbers*, p. 524.

25 Peter Paret, *Understanding War: Essays on Clausewitz and the History of Military Power* (New Jersey: Princeton University Press, 1993), p. 137.

26 Dylan Evans, *Emotion: A Very Short Introduction* (Oxford: Oxford University Press, 2001), Preface.

27 Gray, *Black Mass*, p. 87.

theory that aims at scientific rigour should not only not ignore, but should actively seek to incorporate.

Events of the day

In addition to the rich vein of intellectual influences, we can also attribute Clausewitz's emphasis on such forces of passion and hatred in large part to the turbulent events which he lived through.

It is first useful to contrast the Napoleonic period, which dominated Clausewitz's professional career, with the preceding form of warfare which he partly experienced in the wars against Revolutionary France. Supposedly reflecting the civilised spirit of the age, eighteenth-century warfare was typified by its limited nature. Most armies approximated to a 'state within a state' in which harsh discipline was upheld to ensure troops could perform intricate evolutions on the battlefield, withstand enemy fire at point-blank range and not flee from the battlefield entirely. In this respect, the emotions of the troops had to be reigned in.[28] Churchill once quipped that in these armies, 'bad temper was not often permitted to intrude'.[29] When Clausewitz went to war in 1793 he did so as part of just such an army. Yet, before long, much of this form of war would be violently swept away.

The storming of the Bastille in France on 14 July 1789 heralded a new era in which the impulses of 'the people' began to affect the behaviour of the state as much as the interests of 'monarchs, prelates, and aristocrats'.[30] Europe would soon ensure a return to established principles of diplomacy after the Congress of Vienna, but the popular forces unleashed by the Revolution were irreversible. Clausewitz clearly recognised the importance of these changes and their enormous consequence in the realm of military affairs. The transfer of sovereignty from the monarch to the nation radically altered the underlying meaning of the monopoly of legitimate violence.[31] That monopoly became the preserve of the representatives of the 'Third Estate' – the masses, the people – led by men fired by an intense belief in the cause of liberty, equality and fraternity as encapsulated in the *Declaration of the Rights of Man and Citizen*. Despite the shift towards 'the age of dictatorship,'[32] with the crowning of Napoleon as Emperor in 1804 and the concomitant decline

28 Bond, *Pursuit*, p. 15.

29 Quoted in Gunther Rothenberg, 'The Age of Napoleon', in Michael Howard, George Andreopoulos, and Mark Shulman (eds), *The Laws of War: Constraints on Warfare in the Western World* (New York: Yale University Press, 1994), p. 86.

30 Tim Blanning, 'Epilogue: The Old Order Transformed 1789–1815', in Euan Cameron (ed.), *Early Modern Europe* (Oxford: Oxford University Press, 1999), p. 345.

31 A concept developed by Max Weber in his *Politics as a Vocation* (1919). Available online at: <http://www.ne.jp/asahi/moriyuki/abukuma/weber/lecture/politics_vocation.html>, retrieved 6 June 2008.

32 Alfred Cobban, *A History of Modern France – Volume II: 1799–1871* (London: Penguin, 1965).

of popular sovereignty, it was already clear that the age of demagogic politics had arrived. This would have a far-reaching impact on the character of war. As Clausewitz put it, 'in 1793 a force appeared that beggared all imagination. Suddenly war became the business of the people – a people of thirty millions, all of whom considered themselves to be citizens'.[33]

The merging of the 'three great abstractions' of state, nation and people gave rise to forces never witnessed before in European affairs.[34] Supported by the influential thinkers of the Counter-Enlightenment, the powerful ideology of modern 'centrifugal nationalism'[35] was unleashed. It was Napoleon's ability to exploit these new passions that explains his remarkable military successes. He channelled patriotic fervour into the cult of his own person, so that eventually the two became indistinguishable: the future of the Revolution rested on the continued successes of the Emperor. In Clausewitz's words, 'war was returned to the people'.[36]

One might empathise with those Frenchmen who rushed to the defence of *la patrie en danger*, 'fighting as free men to defend freedom'.[37] These citizens-in-arms were injected with a sense of enthusiasm, vigour and 'high pitched morale'[38] almost entirely absent from the professional and largely hired armies which were the dynastic instruments of the old regime. What the revolutionary armies lacked in professionalism they made up for in the vigour of necessity. Although nationalism was not a new concept, it had meant little to the masses who were more concerned with eking out an existence within their locality, their *pays*.[39] After the Revolution, it became evident to these people that their new-found freedoms, however real, depended on the survival of the nation. The Republic gave birth to an army of patriots willing to sacrifice themselves for the common good.[40]

These were historically unprecedented times in which military expansion rode on the crest of the wave of popular enthusiasm. As Clausewitz observed,

> The revolutionary methods of the French had attacked the traditional ways of warfare like acid; they had freed the terrible element of war from its ancient diplomatic and economic bonds. Now war stepped forth in all its raw violence, dragging along an immense accumulation of power; and nothing met the eye but

33 *On War*, VIII, 3B, p. 715.

34 Blanning, 'Epilogue', p. 372.

35 Michael Howard, *The Causes of War* (Cambridge, Massachusetts: Harvard University Press, 1983), p. 161.

36 Clausewitz, 'On the Life and Character of Scharnhorst', p. 102.

37 Michael Howard, *War in European History* (Oxford: Oxford University Press, 1976), p. 79.

38 Rapoport, 'Editors Introduction', in Clausewitz, Carl von, *On War*, trans. J. J. Graham and ed. F. N. Maude, (London: Penguin, 1982), p. 21.

39 Howard, *Causes*, p. 159.

40 Paul Kennedy, *The Rise and Fall of the Great Powers* (London: Fontana Press, 1989), p. 158.

> the ruins of the traditional art of war on the one hand, and incredible successes on the other.[41]

When Clausewitz refers to the people's new found interest in war, it is not in some indirect sense such as 'public opinion', but largely in terms of their direct experience in battle through the imposition of universal mass conscription. Later, and largely in reaction to the 'danger that France posed to everyone'[42] the peoples' involvement in war increased throughout Europe in the form of armed liberation movements, as famously embodied in the Spanish War, but also in the Tyrolean rebellion after 1809, the actions of Cossacks against Napoleon's retreat from Russia and the extensive civilian forces mobilised to form the Prussian *Landwehr* from March 1813.

Given the incredible brutality of these 'people's wars', as Gray states, 'Not for nothing did Clausewitz … associate violence and passion most closely with 'the people'.[43] These developments certainly allowed him to perceive the extent to which passion can fundamentally transform politics and warfare. This was clearly only one possible – albeit perhaps the most potent – way in which passion materialises in war and quite understandably came to dominate his thinking. Without Clausewitz's exposure to these events, it has to be doubted whether he would have fully incorporated these ideas into his theory or invested them with the importance that he did.

Clausewitz's personal experiences

It remains to note briefly that not only was Clausewitz acutely aware of these new passions that had been unleashed across Europe in a historical sense, but on a number of occasions he came face to face with them, indeed personally lived them, in all their bloody and dramatic reality. Many of the passages concerned with such emotional factors convey a powerfully subjective element, almost as if the author is re-living such events. In the Chapter on 'Danger in War', in which Clausewitz takes us on a journey with a soldier cautiously entering into the thick of battle, there can be little doubt that the soldier is Clausewitz, perhaps recounting his first taste of war as a young teenager against the Revolutionary armies.[44]

Moreover, Clausewitz himself was stirred by such emotions, or what in *On War* he termed 'hostile feelings'. He was, at certain times in his career, enthused with a violent and energetic patriotism, which almost threatened to usurp the

---

41 Carl von Clausewitz, 'On the Life and Character of Scharnhorst', in Carl von Clausewitz, *Historical and Political Writings*, ed. and trans. Peter Paret and Daniel Moran (New Jersey: Princeton University Press, 1992), p. 102.

42 *On War*, VIII, 3B, p. 717.

43 Colin S. Gray, War, Peace and International Relations: An Introduction to Strategic History (Abingdon: Routledge, 2007), p. 246.

44 *On War*, I, 4, p. 132.

measured side of his personality. While in exile, in 1807 he wrote to Marie that, as regards Germany, 'with whips I would stir the lazy animal and teach it to burst the chains with which out of cowardice and fear it permitted itself to be bound'.[45] Such sentiments are also forcefully revealed in his political testament of 1812 – after Frederick William had signed what Clausewitz believed to be a humiliating treaty with Napoleon – in which he states that he would consider himself 'lucky to die gloriously in a noble struggle for the freedom and dignity of the fatherland'.[46]

Clausewitz certainly understood, directly and personally, the way in which war can induce feelings of patriotic fervour, group belonging and intense enmity: his life-long hatred of the French – magnified after the humiliation of 1806 and his being witness to the pitiful state Prussia had been reduced to prior to 1813 – is testament to this.[47] In his essay on 'The Germans and the French', perhaps restraining himself somewhat, Clausewitz writes: 'I will say nothing here of the well-known frivolity and facetiousness of the French'.[48] However much of a detached intellectual he was, Clausewitz was certainly not immune to irrational hatreds and impulsive behaviour.

Explorations

Of Clausewitz's three tendencies, this one of passion is perhaps the most intellectually awkward to swallow: to those from supposedly 'enlightened' or 'civilised' states, it clashes with norms of military behaviour, perceptions of self-righteousness and a progressive belief in the triumph of reason over the passions. Indeed, Coker has argued that this is how Western warriors have distinguished themselves: 'War was the business of "civil" people not enslaved to their passions or pursuing their enemies out of revenge. Western warriors were rational'.[49] While believing that less civilised states and armed groups in other parts of the world are often more in the thrall of aggressive emotion when they go to war, Westerners claim to do so with a clear head, for reasons of state interest and led by the cold calculation of the mind, not the hot blooded command of the heart.

This tendency of the trinity, they might argue, no longer applies to them – war is a rational activity undertaken by responsible leaders elected by a well-informed public – and while chance may still apply, passion has been controlled by universal education and the authority of the intellect. This was a view reinforced during the early phases of the Cold War when strategies for potential nuclear war were

45 Quoted in Paret, *Clausewitz*, p. 129.

46 Carl von Clausewitz, 'Political Declaration', in Clausewitz, *Writings*, p. 291.

47 Parkinson, *Clausewitz*, p. 26.

48 Carl von Clausewitz, 'The Germans and French', in Clausewitz, *Writings*, p. 255.

49 Christopher Coker, *Waging War Without Warriors? The Changing Culture of Military Conflict* (London: Lynne Reiner, 2002), p. 107.

reduced to statistical metrics and mathematical, 'economic conflict' models.[50] Perhaps because the emotional strains of 'thinking about the unthinkable' would be too much for any sane individual to face, the human consequences had to be converted into cold euphemisms such as 'city avoidance'.[51]

Furthermore, Western depictions of brutal conflicts in the Balkans, Africa and other parts of the developing world are often explained away as simply the 'atavistic eruptions of ... incorrigible tribalism', the actions of savages fired by ancient hatreds, who, because of their backward social conditions, are unable to conduct war with the restraint shown by Western states. As Martin Smith argues, the concentration on the irrational in these wars represents more 'a means to explain away the essential unwillingness to go through the laborious task of understanding such wars'.[52] This is not to argue that such deep-seated hatreds are not present, just that a crude rendering shrouds the political contexts and logic of such wars, while setting up unjustified and haughty notions of Western racial and cultural superiority.

A closer examination of such wars reveals the continued workings of rational and strategic behaviour, even where gross crimes against humanity, ethnic cleansing and barbaric acts take place. Sadly, such acts often stem precisely from their political logic. It is the nature of the political issues – such as power struggles shaped by ethnic identity – over which those wars are fought that provides them with much of their brutal character, not mere barbarism. The problem is more an inability on the part of Western analysts, due to ethnocentric blinkers, to deduce the rationales behind such actions. This lack of understanding might be deemed surprising given that during World War II, the strategic bombing of whole cities, which was bound to result in thousands of civilian deaths, was conducted for ostensibly rational strategic reasons by supposedly civilised states.

On this point, and rejecting the myopic 'progressive' Enlightenment perspective, Clausewitz argues that hostile feelings exist regardless of any supposed level of civilisation a society may have reached. 'Savage people' Clausewitz notes 'are ruled by passion, civilised people by the mind'[53] but, crucially, 'even among educated peoples and civilised societies men are often swept away by passion, just as in the Middle Ages poachers chained to stags were carried off into the forest'.[54] Civilised peoples are just as susceptible to these forces and the extent of

50 Colin S. Gray, 'What RAND Hath Wrought', *Foreign Policy*, No. 4, Autumn 1971, p. 111.

51 An array of civilian strategists developed 'scientific' theories based on assumptions of rationality and an abstract 'homo strategicus'. See Lawrence Freedman, *The Evolution of Nuclear Strategy* (Hampshire: Palgrave Macmillan, 2003), p. 231.

52 M. L. R. Smith, 'Strategy in an Age of Low-Intensity Warfare', in Isabelle Duyvesteyn and Jan Angstrom (eds), *Rethinking the Nature of War* (London: Frank Cass, 2005), p. 51.

53 *On War*, I, 1, § 3, p. 84.

54 *On War*, I, 3, p. 124.

their influence will depend on such factors as the importance of the 'conflicting interests' and 'on how long their conflict lasts'.[55] Reason and civilisation, he claims, may limit some of the excesses in war – if 'civilised nations do not put their prisoners to death or devastate cities and countries, it is because intelligence plays a larger part in their methods of warfare'[56] – but cannot radically alter the basic human condition.

Clausewitz on passion

Handel comments that, 'Clausewitz was perhaps the first major strategist to give full recognition to the influence of non-rational elements on warfare'.[57] References to the emotional aspects of war occur frequently throughout *On War* and in much of Clausewitz's other work. Indeed, when contrasted with the generally dry, technical and anodyne nature of Enlightenment military thinkers the originality is stark. At one point, Clausewitz remarks that 'in the dreadful presence of suffering and danger, emotion can easily overwhelm intellectual conviction'.[58] Clausewitz refers to a soldier whose heart is moved to 'awe and pity' by the terrible things he sees in battle and that he cannot pass through the 'layers of increasing intensity of danger without sensing that here ideas are governed by other factors, that the light of reason is refracted in a manner quite different from that which is normal in academic speculation'.[59] While such passages certainly add to the gripping and powerful character of much of the prose, the inclusion of such imagery was intended less for aesthetic effect than for the higher purpose of theoretical precision. Clausewitz was certainly no romantic poet[60] and he was not even writing for public acclaim, but rather for personal intellectual fulfilment and lasting influence.[61]

The inclusion of such imagery might appear to be the personal sensibilities of a passionate soldier creeping into his supposedly sober objective analysis, thus throwing into question its scientific credentials. Yet, his inclusion of emotional factors in theory was, he believed, essential for any 'scientific' treatment of the subject – such factors were for him inescapable elements of the reality of war, not mere side-effects or aspects that the rational observer must see past in order to get to the root of the subject, nor elements belonging to some other transcendent realm as some previous thinkers had held.

---

55 *On War*, I, 1, § 3, p. 85.

56 *On War*, I, 1, § 3, p. 85.

57 Michael I. Handel, *War, Strategy and Intelligence* (London: Routledge, 1989), p. 15.

58 *On War*, I, 3, p. 125.

59 *On War*, I, 4, pp. 132–3.

60 Mike Smith notes how some commentators in US military institutions believe *On War* should be taught as poetry. Smith, 'Low-Intensity Warfare', p. 39.

61 See 'Preface by Marie von Clausewitz to the Posthumous Edition of Her Husband's Works, Including *On War*', in *On War*, pp. 73–5.

References to passion and irrationalism appear in different contexts throughout Clausewitz's work and are utilised for differing illustrative or conceptual purposes. This makes study of these ideas a difficult task for the student of Clausewitz. Passion is closely associated with what Clausewitz terms 'moral forces', but is not exactly coterminous with them; moral forces have a broader application and extend to subjects not properly considered in the context of this tendency, such as courage or the psychology of the commander. These points are intended to emphasise the extent to which it is more or less futile to separate elements of passion from other important subjects which it feeds into and complements, but which are best covered in other sections of this book. Nevertheless, a close examination yields a fairly clear picture of what Clausewitz was intending to convey.

Individual and collective passions

At the level of personal experiences in war this tendency is perhaps fairly uncontroversial and clear. A glance through military diaries from any age will confirm this. Keegan emphasises the importance of these 'sensations and emotions' and how they constitute a powerful 'part of every human being's make-up … They touch too upon some of man's most violent passions; hatred, rage and the urge to kill'.[62] They are permanent and universal. As Michael Dockrill makes clear, the missiles have become much more accurate and protective measures … are more sophisticated, but the human frame has not changed. The feelings and emotions of those exposed to firepower, whether as civilians and combatants, have remained constant throughout history'. [63]

While an individualistic conception of emotion is important to our understanding, for Clausewitz, the forces described are presented less as personal than as collective, social phenomena. For instance, he explains how 'even the most civilised of *peoples* … can be fired with a passionate hatred for each other'.[64] The types of forces described are often referred to as the 'mood of the nation' or, as Clausewitz puts it, the 'temper of the population.[65] Was he right to write of such *collective* passions and, if so, how exactly are they to be conceived?

Mass consciousness is dimly perceived but impossibly abstract.[66] Collective emotion is something more all-pervasive than simply the aggregate of each individual's feelings, the resulting whole of which is perhaps felt by all and outlasts the particular moment. It is something more powerful, yet simultaneously elusive to the observer. It is a shifting, morphing force, the presence of which may be sensed, but that is only truly realised through symbolic action or public

62 John Keegan, *The Face of Battle* (London: Pimlico, 2004), p. 18.

63 Michael Dockrill, 'The Experience of War', in Lawrence Freedman (ed.), *War* (Oxford: Oxford University Press, 1994), p. 9.

64 *On War*, I, 1, § 3, p. 84. Emphasis added.

65 *On War*, III, 3, p. 216.

66 See Bendedict Anderson, *Imagined Communities* (London: Verso, 2006).

expressions of violence, like a rain cloud made up of millions of tiny droplets blown along together and shaped by the wind, which appears to exist as some kind of whole, but only manifests itself as concrete reality when its collective density reaches such a level that it results in a sudden downpour or flash of lightening. In fact, Clausewitz evokes similar imagery with respect to a people's uprising against occupation in which, 'the fog must thicken and form a dark and menacing cloud out of which a bolt of lightening may strike at any time'.[67]

Collective feelings can exist as a kind of independent force, both dependent on the individuals who brought them into being and affecting them in turn. They can be deliberately inculcated by politicians or emerge spontaneously in response to certain events. The example of a mass rally is a pertinent instance of the extremes collective passions can reach. Indeed, many people attest to the great power of 'the crowd' and how it can induce emotions that otherwise would not exist. Ehrenreich notes how the crowd 'leaves mundane things behind and transmutes itself into a new kind of being, larger than the sum of its parts, more powerful than any single individual'.[68]

The concept is however temporally and spatially flexible and can relate to the sense of brotherhood felt by a small terrorist organisation or extend to the general sentiments of an entire nation. Additionally, it must be added that such a collective concept can only refer to a generalised aggregate, the dominating characteristics of the whole, which will be composed of a vast range of individual and sometimes conflicting emotions on closer inspection. For instance, the enthusiasms for war displayed in August 1914 were not shared by all, nevertheless it is the scope of the enthusiasm that was historically important.[69] As Howard attests in relation to such emotional forces during World War I, 'trends like these are difficult and sometimes impossible to document, but no study of the war can be complete which does not take them into account'.[70]

Passion and the use of force

Clausewitz strongly identifies this tendency (often shortened to the synonym, 'violence') with the physical use of force, but we should be careful not to confuse the two. Here, Clausewitz is concerned with violence as a *motive force*, not in the physical use of the term.[71] Yet, the former is inescapably intertwined with the latter to the extent that emotions are invariably aroused where physical force is a

67 *On War*, VI, 6, p. 581.

68 Ehrenreich, *Blood Rites*, p. 16.

69 Bond, *Pursuit*, p. 101.

70 Michael Howard, *Studies in War and Peace* (London: Temple Smith, 1970), pp. 103–4.

71 See 'The Primacy of Policy and the "Trinity" in Clausewitz's Mature Thought', in Hew Strachan and Andreas Herberg-Rothe (eds), *Clausewitz in the Twenty-First Century* (Oxford: Oxford University Press, 2007), p. 82.

feature of the relationship between humans, whether threatened or actually used. In Clausewitz's work a definite connection between the act of force and emotion is revealed and they are presented as almost inseparable: at one point Clausewitz notes that, 'Essentially, combat is an expression of *hostile feelings*'.[72]

In Book 1, Chapter 1 of *On War*, Clausewitz discusses the first interaction to the extreme – 'the Maximum Use of Force' – which can be summarised as the logic of force outdoing opposing force in the abstract: if one side uses force unreservedly, the other will be compelled to follow suit or risk being rendered powerless.[73] This rise to extremes is not necessarily determined by hostile emotions, nevertheless Clausewitz holds that emotions cannot fail to become involved and they then often serve to intensify the process. War, as an act of force, has an inherent escalatory logic, compelled significantly by, as Daniel Moran has put it, the 'violent, irrational emotions that the use of force could inspire'.[74] A crucial feature of this reciprocal action is the concept of hostile feelings.

Clausewitz notes a definite link between these emotional aspects and the physical act of fighting by stating that it would be a fallacy 'to conceive of war as capable of ridding itself of passion', and to do so would imply that fighting would be unnecessary: an algebraical comparison would suffice to settle the matter if war were purely a matter of the intellect. War can never simply be a rational act: 'if war is an act of force, the emotions cannot fail to be involved'. The idea is also expressed in his use of the phrase, 'the *impulse* to destroy the enemy,'[75] and later in Book 1 he refers to the, 'brute discharge of hatred and enmity of a physical encounter'.[76]

From Achilles' all-consuming revenge and relentless fury at Hector for the death of Patroclus, these kinds of forces, described by Clausewitz as a 'flaring-up of mutual rage',[77] have been powerfully recorded throughout history. One recent book turns to Clausewitz in order to express the idea. In Chris Bellamy's comprehensive history of the gruesome experience of Soviet and Axis forces on the Eastern Front during World War II, he stresses the role such violent emotional forces played in pushing the fighting perilously close to Clausewitz's abstract concept of absolute war: 'Violence by one side bred violence on the other. Soviet forces breaking into Germany in 1945 were spurred on by exhortations to exact revenge'.[78] He goes on to detail the way in which the notable absence of legal restrictions gave freer rein to these destructive and violent feelings between the two countries.[79]

72 *On War*, II, 2, p. 158.

73 *On War*, I, 1, § 3, p. 84.

74 Daniel Moran, 'The Instrument: Clausewitz on Aims and Objectives in War', in Strachan and Herberg-Rothe, *Clausewitz*, p. 91.

75 *On War*, I, 1, § 3, pp. 84–5.

76 *On War*, I, 2, p. 108.

77 *On War*, II, 2, p. 159.

78 Chris Bellamy, *Absolute War* (London: Macmillan, 2007), p. 19.

79 Bellamy, *Absolute War*, pp. 20–31.

A similar dynamic is highlighted by Ferguson in his book, *The War of the World*. In a section revealingly entitled 'Hatred in the Trenches', which explores the failure of socialist hopes of international proletarian solidarity superseding national loyalties during World War I, Ferguson describes how 'as the war went on, mutual hatred grew, expunging the common origins and predicaments of the combatants'. Many began to regard the opposing soldiers as 'a sort of vermin like plague-rat', hate replaced fear, and, for some, killing became a cathartic, even pleasurable, experience.[80] That such dynamics as explained by Ferguson were included in Clausewitz's meaning of this tendency is clear: 'Even where there is no national hatred and no animosity to start with, the fighting itself will stir up hostile feelings: violence committed on superior orders will stir up the desire for revenge and retaliation against the perpetrator'.[81]

Extreme hatred can cause mindless killing and acts of barbarity to become a matter of course for those involved. So, Achilles, gripped by rage, desecrates Hector's corpse, and he even 'outrages the senseless clay in all his fury'.[82] Sadly human history records far too many instances of such brutality. These examples are certainly some of the most obvious ways in which passions are aroused in war. 'Hostile feelings' will generally be aroused where violence occurs in human social relations.

Sources of passion and irrationality

As Clausewitz himself notes, it would be pedantic to try to outline all the various ways in which emotional, psychological and moral factors impact on war. Nevertheless, it is perhaps useful, drawing on some ideas Clausewitz discusses, to consider some of the most notable sources of passion in relation to war other than those associated with the act of force itself as described above. The different conditions created by and characterising any particular war accounts for the differing extents to which emotions are aroused or irrational influences come to dominate the conduct of war: levels of hostility can fluctuate from one conflict to the next or even multiple times within the same war.[83]

Of course, in some wars emotions are actively roused by the leadership in order to drum up enthusiasm or attain support.[84] The deliberate mobilisation of hostile feelings is common practice: fear may be encouraged by exaggerating the threat posed by the enemy, while 'hate training' and propaganda may be employed

---

80 Niall Ferguson, *The War of the World: History's Age of Hatred* (London: Penguin, 2006), pp. 123–30.

81 *On War*, II, 2, p. 159.

82 Homer, *The Iliad*, p. 590.

83 Antulio J. Echevarria II, *Clausewitz and Contemporary War* (Oxford: Oxford University Press, 2007), p. 73.

84 See Colin McInnes, *Spectator-Sport Warfare: The West and Contemporary Conflict* (Colorado: Lynne Rienner, 2002), p. 63.

or the media utilised to portray 'the other side as vermin, insects, dogs, and other noisome creatures'.[85] Such practices often result in awful atrocities – the enemy is perceived as sub-human and therefore exceptions to norms are considered justified, even necessary.

The length of hostilities can have a great impact on the play of emotional forces. Clausewitz explains how such ceaseless activity would 'arouse men's feelings and inject them with more passion and elemental strength'.[86] Of course, time may, in some circumstances, have an opposite effect whereby the vigour, passion and enthusiasm displayed in the early parts of a war are dampened by the gruesome reality of the ordeal. Jomini points to this phenomenon in relation to wars, such as those of the Revolution, in which human passions were 'excited in a temporary paroxysm, of less duration as it is the more violent'. In these cases, 'Time is the true remedy' as the 'storms soon pass away and reason resumes her sway'.[87] Clausewitz similarly notes how wars of national liberation may 'start out full of vigor and enthusiasm' but if many soldiers and civilians are killed, wounded, or taken prisoner then 'such defeats would soon dampen its ardour'.[88]

Linked to the idea of time is the factor of the cumulative impact of lives lost and sacrifices already made, which often serves to justify continued fighting for the sake of those already fallen and fuels anger and animosity towards the enemy. Fighting in such cases tends to be suffused with notions of revenge and retribution. This constitutes a major instance of the way in which the course of war itself can cause military dynamics to escape policy control. The emotional imperative to achieve victory for the sake of those already fallen may inspire continued resistance where there is little rational hope of success. This was a factor which contributed to the slaughter of World War I. As Dyer notes, when 'one in three of the male population … has been killed or wounded … when the people's willingness to go on making sacrifices has been sustained by hate propaganda that depicts the war as a moral crusade against fathomless evil – then governments cannot just stop the fighting'.[89]

Perhaps the most potent emotion in war is quite simply fear. There is nothing like fear to overwhelm reason and induce behaviour which people ordinarily would not think themselves capable of. Within rebel groups, fear may manifest itself in paranoia regarding spies, turncoats or fifth columnists, which can induce internal dislocation and indiscriminate purges. Indeed, elements of the Algerian

85 Michael Ignatieff, *The Warrior's Honor: Ethnic War and the Modern Conscience* (London: Vintage, 1999), p. 56.

86 *On War*, I, 1, § 14, p. 93.

87 Antoine Henri de Jomini, *The Art of War*, trans. G. H. Mendell and W. P. Craighill (Texas: El Paso Norte Press, 2005), p. 20.

88 *On War*, VI, 26, pp. 582–3.

89 Dyer, *Lethal Custom*, pp. 265–6.

F.L.N. – an insurgent organisation fighting for independence from France in the 1950s and 60s – almost destroyed themselves this way.[90]

The effect of fear is particularly noticeable in cases of units, armies or even whole nations that face probable defeat and particularly where capitulation or retreat is not an option. In such situations, desperation can generate great shows of strength fuelled by passion, much as an animal is most dangerous when cornered or on the verge of death. In the section of *On War* on the 'culminating point of victory', Clausewitz notes how the commander must assess whether 'the burning pain of the injury he has dealt will make the enemy collapse with exhaustion or, like a wounded bull, arouse his rage'.[91] On this phenomenon, the seventeenth-century jurist Hugo Grotius noted that, 'there is the boldness of despair to be apprehended from a vanquished enemy, dangerous as the bite of a ferocious animal in the pangs of death'.[92] At the tactical level, a classic piece of military wisdom advises always to leave your enemy a 'golden bridge' to avoid the ferocity with which a cornered group is likely to resist, because in a battle for survival 'anything goes'. Tellingly, in his *Guerrilla Warfare*, Che Guevara refers to the 'encirclement face' displayed by surrounded troops.[93] Some of military history's most notable battles occurred under such circumstances, from the Spartans' 'glorious annihilation' at Thermopylae in 480, to the British stand at Rorke's Drift in 1879, to the Russian defence of Stalingrad over the Winter of 1942–3.

The factor of fear is closely related to the types of wars that are, as Clausewitz puts it, 'a struggle for political existence'.[94] Clausewitz states that, 'like a drowning man who will clutch instinctively at a straw, it is a natural law of the moral world that a nation that finds itself on the brink of an abyss will try to save itself by any means'.[95] Elsewhere he notes that, 'No nation has ever responded to repression by another with anything except hatred and enmity'.[96] Similarly, Jomini warns about engaging in such 'national wars'.[97] He advises that to succeed it is necessary to avoid giving the impression that independence is threatened and to 'calm the popular passions in every possible way'.[98] Like Jomini, Clausewitz lived through a period in which countries threatened by emasculation under the Napoleonic Empire fought brutal campaigns of liberation. The greater spirit and vigour of

90 See Horne, *Savage War of Peace*, pp. 258–61 and pp. 322–5.

91 *On War*, VII, 22, pp. 692–3.

92 Quoted in Paddy Ashdown, *Swords and Ploughshares: Bringing Peace to the 21st Century* (London: Weidenfeld & Nicolson, 2007), p. 43.

93 Ernesto 'Che' Guevara, *Guerrilla Warfare* (London: Souvenir Press, 2003), p. 63.

94 *On War*, I, 2, p. 108.

95 Quoted in Handel, *Masters*, p. 198.

96 Carl von Clausewitz, 'Notes on History and Politics', in Clausewitz, *Writings*, p. 270.

97 Jomini, *Art of War*, p. 27.

98 Jomini, *Art of War*, p. 25.

citizen-soldiers fighting for independence has been apparent since Herodotus proclaimed it to be a crucial factor in the Greek victory over Persia.[99]

Other sources of passion and irrationalism may derive from religious belief or superstition, the impact of which can be varied and complex: it can be a potent cause of war, impinge on how it is conducted or shape the conventions under which it is fought. Although wars are rarely purely about religious matters, wars fought over religious differences have been common in human history. Many other wars, not ostensibly 'about' religion have been greatly influenced, or rendered more intractable, by matters of faith, such as in Northern Ireland, Bosnia and the Levant in the twentieth century. Faith might induce hope of salvation where there is none or encourage actions that are strategically crippling, while religious conviction can greatly contribute to morale. Medieval warfare was particularly influenced by such notions, where divine providence was believed to be the prime determinant of success. Religious influences, while often encouraging excesses bred of the need to stamp out apostasy or forcefully convert unbelievers, may also constitute powerful forces discouraging violence and encouraging restraint, pity or mercy during war.

Closely related is the influence of secular 'quasi-religious' or millenarian political ideologies.[100] Jomini noted that wars fought for political dogmas 'are most deplorable … [as] they enlist the worst passions and become vindictive, cruel, and terrible'.[101] It was the passionate utopianism of the Revolutionaries, which Clausewitz recognised as a major factor enabling the rag-tag French armies to first successfully face the professional armies of the European monarchies and then to move 'with such confidence and certainty' during the wars of the following years.[102] The depth of feeling associated with such wars can lead to impossible dreams, hubris and ultimately, nemesis as ideological orthodoxy is often maintained at the expense of strategic sense.

Finally, we should not underestimate the extent to which the state of existing technology can impact on the emotional aspects of war. Reduced proximity between belligerents engendered by rapid advances in weapons development – principally over the last 150 years – has raised the possibility of war fought with a minimum of emotion. Modern 'push-button warfare' is said to be more impersonal, sterile and thus less susceptible to being influenced by the intense emotions of close combat. The apotheosis of this form of warfare came in 1999 with Nato's air war over Kosovo, fought at 15,000 feet and with zero alliance casualties.[103] Yet,

99 Herodotus, *The Histories*, trans. Aubrey de Sélincourt (London: Penguin, 1996), p. 307.

100 Martin Wight, *Power Politics* (London: Pelican, 1979), pp. 139–40.

101 Jomini, *The Art of War*, p. 19.

102 *On War*, VIII, 3B, p. 716.

103 'Clausewitz would have called Kosovo a cabinet war … it was fought by no more than 1,500 NATO airmen, and the elite specialists of the Serbian air defense'. Michael Ignatieff, *Virtual War: Kosovo and Beyond* (London: Vintage, 2001), p. 111.

there are a number of reasons for caution with regard to the supposed limitation of emotions in modern technological warfare.

First, Kaplan suggests that the increased distance between a violent act and the perpetrator might actually expand 'the scope of impersonal viciousness'.[104] Second, technology is not the only determinant of emotions; the political context is vital. One can perhaps expect a US pilot, in the aftermath of the 9/11 terrorist attacks, to display a different emotional disposition towards al-Qaeda terrorists than to Serbian soldiers. Third, even if passion is less evident today, it still exerts considerable influence in more complex and ephemeral forms, such as through public reactions to civilian casualties. We must also remember that war is still decidedly real for those on whom the bombs fall. The reactions of such people are crucial, particularly when most modern Western wars are fought to shape political conditions in the target state. Finally, for all the discussion of 'risk-transfer war'[105] and the increasing use of automated drones, the last decade has seen Western forces engaged consistently in intensely up-close and personal forms of combat. In counterinsurgency warfare, there is no reliable substitute for large numbers of boots on the ground.

There are of course countless other factors that may account for the prevalence of emotional factors in war, but what these examples reveal is that our conception need not be confined to the effect of passion leading to brutalities or extremes in war; it may also result in mistaken decisions, misspent energies, ill-considered interventions, public opposition and so forth. There are more subtle ways in which actors in war display emotional, irrational behaviour, and which might constitute important variables for understanding any particular conflict. To explore this concept further it is useful to consider the ways in which this tendency pertains to specific groups in war: the political leadership, armed forces and the people.

Passion and politicians

In the context of the trinity, the types of forces we have described seemingly threaten to alter, 'if not to overwhelm'[106] the formulation of sensible and reasonable policy or the application of appropriate means. In this sense, strong political leadership is often required to keep popular emotions under control.[107] For instance, the political leadership might have to resist popular demands to wreak vengeance on an enemy, where such a course would not make strategic sense.

---

104 Kaplan, *Warrior Politics*, pp. 11–12.

105 Martin Shaw, *The New Western Way of War: Risk-Transfer War and its Crisis in Iraq* (Cambridge: Polity, 2005).

106 Moran, 'The Instrument', p. 91.

107 Jan Willem Honig, 'Strategy in a Post-Clausewitzian Setting' in Gert de Nooy (ed.), *The Clausewitzian Dictum and the Future of Western Military Strategy* (The Hague: Kluwer Law International, 1997), p. 113.

This tendency represents something that at times policy may be hard put to control and indeed is often seemingly eclipsed by it.[108] Clausewitz notes that, 'Between two peoples … there can be such tensions, such a mass of inflammable material'[109] that the war can develop in a way disproportionate to original political interests. Policy must understand that when it employs force, the emotional reactions of their own forces and of others are highly unstable and unpredictable. Policy might wish to fight a limited war, but events might lead to a different outcome. For instance, the use of force might contribute to 'political goal inflation' as the public demands objectives which reflect the sacrifices already made.[110]

Not only does policy attempt to manage such passions, but there is every reason to believe that it can be overwhelmed by them itself: as Creveld notes, 'Nothing would be more preposterous than to think that, just because some people wield power, they act like calculating machines that are unswayed by passions'.[111] The extent to which policy can remain immune from these forces will depend on such factors as the psychological attributes of the leadership and the character of the war-making institutions. Frederick the Great is a good example of a commander who treated his wars as games of power politics, conducted through the calculated application of his resources and considered manoeuvring of his forces in pursuit of clear and achievable aims.[112] About Frederick, Clausewitz noted that with regard to his political objectives: 'neither vanity, ambition, not vindictiveness could move him from his course; and it was this course alone that brought him success'.[113]

Yet, of course, not all leaders and commanders show the type of controlled reason of Frederick, and those responsible for crafting policy can easily be swayed by the types of forces we have described above. Clausewitz refers to these as emotions that are not so much stimulated by the nature of combat itself, but related to it more obliquely: he mentions – no doubt with Napoleon in mind – 'ambition, love of power, enthusiasms of all kinds'.[114] Napoleon perhaps represents the opposite of Frederick in this respect: although he had a rational side, 'the realistic ruler was dominated by the romantic conqueror'.[115] So, likewise, the failure of Athens occurred, according to Thucydides, because they had 'been seized by a mad passion to possess that which is out of reach' and were 'daring beyond their power, bold beyond their judgement'.[116] Also, associated with the type of hubris

108 *On War*, I, 1, § 25, pp. 99–100.
109 *On War*, I, 1, § 11, p. 90.
110 Honig, 'Strategy', p. 117.
111 Creveld, *Transformation*, p. 157.
112 Bond, *Pursuit*, p. 23.
113 *On War*, III, 1, p. 210.
114 *On War*, II, 2, p. 159.
115 Clark, Kenneth, *Civilisation: A Personal View* (St Ives: John Murray, 2005), p. 213.
116 Thucydides quoted in Coker, *Waging War*, p. 20.

which led Athens into its reckless Sicilian expedition,[117] the impact of overreaching ambition, utopian ideology or blinding faith can cause policymakers to make irrational decisions that are not suited to political conditions or available means.

Passion and the military

That 'the military', broadly defined, does not at first sight warrant direct association with this tendency might appear counterintuitive: surely it is soldiers who are most prone to passionate hatred of the enemy. However, there are important reasons why Clausewitz may have shied away from such a conclusion, derived from a peculiar way of thinking that has become deeply embedded in the culture of military establishments throughout history.

The notion of the military constituting a passionless realm is evident in the tradition of 'de-personalising' soldiers – as Clausewitz put it, turning them into 'automata' – in order that they are simply capable of carrying out orders and to help them 'avert the onset of fear or, worse, of panic and to perceive a face of battle which, if not familiar, and certainly not friendly, need not, in the event, prove wholly petrifying'.[118] As noted, in the age of Frederick the Great armies had to be capable of withstanding fire at point blank range. Moreover, they were generally 'composed of the slime of the nation'[119] and foreign mercenaries, so a great effort had to be made simply to avert mass desertion.

The extremely harsh discipline characteristic of the Prussian army in Clausewitz's time, particularly prior to the reform period, would have been common to most soldiers throughout Europe. The army was an institution built around the need to restrict the play of blind emotion. It is understandable why Clausewitz perhaps refrained from a direct association of passion with the military given the practical requirement for strongly disciplined and controlled armies, but also in the objective sense that such control has been a prominent feature of armed forces throughout history.

Yet, the issue is more complex than this and no rigid distinction is actually implied in Clausewitz's conception. This becomes clear when one considers the nature of military institutions as they developed during the Napoleonic era. Clausewitz witnessed the emergence of the phenomenon of the 'nation in arms' where recognisable distinctions between 'the people' and the army broke down: civilians were soldiers, soldiers civilians. Thus, when Clausewitz speaks of the people, we should not be misled into conceiving of a neat sociological distinction that might apply today, with small professional armies isolated from wider society.

---

117 Thucydides, *Peloponnesian War*, pp. 414–29. Nicias warns the Athenian assembly against the expedition: 'Do not … indulge in hopeless passions for what is not there. Remember that success comes from foresight and not much is gained simply by wishing for it', p. 417.

118 Keegan, *Face of Battle*, pp. 21–2.

119 John A. Lynn, 'States in Conflict', in Parker, *Cambridge History*, p. 180.

The emotions of the people were embodied in the military instrument in the form of the civilian in arms.

Furthermore, surely all soldiers, statesmen and so on, by necessity, are of 'the people' – they are not an alien breed. To claim they are not susceptible to similar emotional forces attributed to 'the masses' is mistaken. This tendency is concerned with the human element in war; all actors in war are human and will display elements of this tendency to some degree. Just as all individuals are incapable of living purely by the dictates of reason – even Stoics as indomitable as Marcus Aurelius or Seneca[120] – so all groups of society, be they ordinary citizens, soldiers, politicians or commanders, can be swept away by passion.

Passion and 'the people'

So, despite having established that the play of emotion in war is not confined to any one element of society – as a narrow reading of the secondary trinity might suggest – Clausewitz nevertheless clearly wanted to stress the extent to which this tendency is powerfully expressed through the people, in its more specific sense of, 'the masses'. The role of populations in war has varied significantly throughout history. As noted, the people were very much separated from the armies of the eighteenth century, composed as they were largely of mercenaries. The rest of the populace was primarily expected to pay taxes and had next to nothing to do with war, except maybe through small segments of middle-class opinion or when peasants found themselves in the path of a marching army. In other times, preceding and post-dating that era, the passions of the people have greatly influenced the conduct of war. For instance, during the Thirty Years War civilians were deeply involved, either as rampaging murdering mobs or as their countless victims.[121]

The emotional predispositions and reactions of the people to uses of force can have crucial political implications, particularly if those emotions are converted into armed resistance of various kinds, but also in the indirect sense of the impact of diminished legitimacy at home and abroad, declining popular support or mass protest. Populations may also be a powerful force either pushing leaders into war or appealing for restraint. Their impact in these respects has gained increasing importance with the spread of liberal-democratic states and mass-media – although as Gray has pointed out, the 'strange idea that dictators can ignore domestic constituencies needs to be quashed once and for ever'.[122] As we stated in Chapter 4, political power is comprised of much more than material military prowess and the effective use of force should be geared to prevailing political conditions. In this

120 As Nancy Sherman notes, 'there are few to no bona fide sages, whether in ancient or modern times, however earnest any given individuals aspirations'. Sherman, *Stoic Warriors*, p. 10.

121 See Victoria Wedgwood, *The Thirty Years War* (New York: New York Review Books, 2005).

122 Gray, 'Clausewitz Rules', p. 172.

respect, securing and maintaining the support of various constituencies of popular sentiment may constitute the difference between success and failure in war: as Jonathan Swift once remarked, 'In war, opinion is nine parts in ten'.[123] Clausewitz was well aware of this, even if its relative importance has grown in recent times: he noted how the centre of gravity – 'the hub of all power and movement on which everything depends' – might in some cases equate to 'the personalities of the leaders and public opinion'.[124]

This is particularly evident, as Clausewitz recognises, in situations of popular uprising and insurgency. In order to offset their material weakness against heavily armed state militaries, insurgents will actively seek to shape the emotions and win the allegiance (or acquiescence) of the population on which they rely both materially and politically. As the axiom has it, 'the population is the prize'. Likewise, counterinsurgent forces must be mindful not to alienate the population through indiscriminate reprisals, collective punishment or the use of disproportionate measures. Insurgents will often seek to induce just such a response, as well as actively seeking to instil fear in segments of the population knowing that 'the worse the better' because it is generally the government that is held responsible by civilians for a failure to protect them, thus draining the latter's all-important legitimacy still further.[125] The fears and the emotions of the population will often determine the outcome more than any material preponderance of force.

A particularly modern manifestation of the impact of popular passions on war has been the rise of pressure from populations in Western states to intervene militarily in the affairs of other states for humanitarian reasons. With the growth of human rights discourse and a global mass-media presenting round-the-clock coverage of conflict and suffering in far-away places, the demand for governments to 'do something' has greatly increased (albeit highly selectively).[126] Such pressures lay behind the succession of interventions by Western states after the end of the Cold War. The outcomes of such interventions were extremely mixed, but the point here is the extent to which the initial engagement, the method of intervention and levels of commitment to the mission in the aftermath have all been crucially shaped by popular opinion – even if it arouses the emotions of the people only 'in the intense but shallow way that sports do'.[127] Often the impulse to do something is highly emotional, in the form of moral outrage, guilt and compassion and rarely is sufficient account taken of the realties and difficulties on the ground. Also, the same emotions that inspire intervention often end up pushing for withdrawal or precipitate 'exit strategies' as 'compassion fatigue'[128] sets in,

---

123 Quoted in Horne, *Savage War of Peace*, p. 231.

124 *On War*, VIII, 4, p. 720.

125 Robert Taber, *The War of the Flea: The Classic Study of Guerrilla Warfare* (Washington, DC: Potomac Books, 2002), p. 6.

126 Ignatieff, *Warrior's Honor*, p. 97.

127 McInnes, *Spectator-Sport Warfare*, p. 55.

128 McInnes, *Spectator-Sport Warfare*, p. 97.

concerns over casualties increase and electorates invariably demand attention on domestic issues.

Implications

Emotion: all bad?

Many of the examples discussed in this chapter have concentrated on the overwhelmingly nefarious types of emotion displayed in war or that are at least considered disagreeable in everyday human relations. These are indeed the types of emotions which Clausewitz emphasises in the trinity itself – violence, hatred and enmity – and they reappear when this topic is brought up in other parts of the work. Yet, there are some indications that Clausewitz held a more balanced view with regard to the types of emotion that might impinge on war.

Clausewitz notes that, 'In considering emotions that have been aroused by hostility and danger as being peculiar to war, we do not mean to exclude all others that accompany man throughout his life' because in war, 'a wide variety of passions, good and bad, will arise on all sides. Envy and generosity, pride and humility, wrath and compassion – *all may appear as effective forces in this great drama*'.[129] So, a more accurate conception of this tendency would perhaps view passion as neither something wholly good or bad. While emotion might certainly be a potent destructive and murderous force that often causes war to rise to extremes, there is certainly another side which is perhaps less obvious.

Humans are of course capable of a wide variety of emotions which we could describe as good, righteous or commendable. It might be thought that such emotions are likely to be absent in war, where fighting, fear and death cause everything to be pervaded by revenge, hate and anger. Yet a closer analysis reveals some interesting observations in the sense that war often displays the full panoply of human emotions. Also, the discussion so far has concentrated more on the emotions apparent *between* belligerents, but a full consideration must account for the emotions *within* armed forces, where often quite different emotional forces rule.

Intense love for one's comrades within armed forces is a prevalent feature of military history. Often this might translate into soldiers wishing it had been them that died rather than their 'buddies', acts of astounding self-sacrifice, the way in which commanders often come to be seen as surrogate fathers (as Clausewitz experienced with respect to Scharnhorst) or the common military maxim never to leave a fallen comrade behind. It is in this sense that Coker can claim that militaries, perhaps somewhat counterintuitively, have altruism at their heart. He notes the intersubjective bonds of honour that exist between warriors and the

129 *On War*, II, 2, p. 160. Emphasis added.

way they form a cohesive guild, sometimes even a fraternity.[130] Some of the most admirable emotions are revealed in this respect: courage, duty, loyalty, truthfulness and community spirit.

Also, a number of historical studies have revealed that in many wars there have developed far more complex relationships between combatants than may be imagined, which go beyond simplistic assumptions of mutual hatred. Indeed, in many cases combatants deny feeling any particular animosity towards the enemy, going as far to even feel a certain measure of intimacy, respect or empathy based on a shared understanding of their common plight.[131] Bourke quotes a telling passage from a World War I soldier's diary that reveals the tension between the differing emotions discussed here:

> On the one side you have all the signs of excessive hate and unbridled passion that shows the innate madness that still lurks in the human soul. On the other you have all the signs of unselfish devotion and kindliness of spirit, even towards the man you have just struck in your hate.[132]

Famously, during that war, a curious phenomenon was the 'live and let live' philosophy that developed in certain sectors of the trenches on the Western Front despite the overwhelming context of 'kill or be killed'. Soldiers came to agreements not to fire if the other side would also desist. There were also widespread instances of fraternisation, as famously embodied in the 'Christmas myth'. Also interesting are the many instances of soldiers who never fired at the enemy, even in extreme circumstances, where their own lives would be in greater danger as a result. Bourke quotes a Canadian military instructor who stated that, far from concerns about trigger-happiness, 'The problem is not to *stop* fire, but to start it'.[133] Studies conducted during World War II revealed a high percentage – sometimes over 75 per cent – of soldiers in combat units who would not fire at the enemy. It is far from clear exactly what accounts for this phenomenon, yet one cannot but suspect that basic emotions of pity, mercy and humanity play a part.

Ambiguity

It might be assumed that this tendency should be understood unequivocally as a force pushing war to extremes. Where it is powerfully present, particularly as embodied in the people's involvement in war, violent emotions can feed into a process of escalation, and we have seen how Clausewitz associated them with the 'maximum use of force'. His discussion is certainly biased towards a conception

130 Coker, *Waging War*, p. 31.
131 Bourke, *Intimate History*, p. 141.
132 Bourke, *Intimate History*, p. 141.
133 Bourke, *Intimate History*, p. 74.

which sees emotion in war as a generally intensifying and escalatory force. The history of war would certainly support such a conception.

Yet, this is not the complete picture. Not only should this tendency not be conceived as entirely negative, but it can also potentially constitute a limiting force on war, for instance as embodied in popular pressures to limit the violence in war, collateral damage and even enemy military casualties.[134] Also, popular sentiment may, far from constituting a force propelling states into war, be a powerful force restraining politicians from the resort to arms. Anti-war and pacifist sentiment grew exponentially during the course of the twentieth century, primarily following the experience of two enormously costly world wars and the prospect of the end of the human race in a third. This was perhaps an aspect of war Clausewitz could not fully foresee given the conditions of his age and its prevailing social and cultural attitudes. Although there have always been people passionately opposed to war throughout history, the historical norm has largely been one of acceptance of war as 'part of the natural order of things', even as a virtuous, commendable activity.[135] The modern rise of anti-war feeling and its specific manifestation in relation to particular wars represents an important factor of this tendency exercising a potentially limiting affect on war.

Strategically neutral

Finally, for Clausewitz, from a subjective strategic perspective, emotional factors are not depicted in an especially negative or positive light: as he notes, for the attainment of some ends 'they will have to be stimulated rather than held back'.[136] Whether emotions constitute a force to be exploited or a genie to be put back in the bottle will depend on circumstances and the specific objectives in view. In some instances passion may greatly aid progress toward objectives in the form of strengthened morale or the support of the people for the cause. The courage instilled in a force inspired by religious devotion or desperate fear might enable it to destroy an enemy twice its size. Also, as we have seen, the positive emotions to be found in war can often constitute powerful factors promoting group cohesion, altruism and self-sacrifice amongst comrades, all of which can enhance military effectiveness.

Conversely, passion can be a highly detrimental force in war. Excessive anger can lead to irrationalism and is almost impervious to control. Seneca considers whether anger might be advantageous to adopt in war – because it 'rouses and spurs on the mind. Without it, courage can achieve nothing magnificent' – but concludes that such a course would be mistaken because 'it is easier to exclude the forces of

134 During the Gulf War, and largely due to the impact of mass media, publics 'became sensitive not only to the losses of their own armed forces, but even to those inflicted on the enemy'. Howard, *War in European History*, p. 142.

135 Howard, *Reinvention of War*, p. 13.

136 *On War*, I, 1, § 25, p. 100.

ruin than to govern them … for once they have established possession, they prove to be more powerful than their governor'.[137] Also, emotion can contribute to the formulation of unrealistic policies or cause military objectives to be set for which the means do not exist. Emotional impulses can encourage the use of unrestrained force which may prove inimical to stated objectives, for instance in encouraging greater resistance from the enemy or by undermining popular support.

While Clausewitz would have sympathised with Seneca's injunctions, he often stresses the need for a delicate balance of emotions, not their complete suppression. This was the case with his notion of genius, which he believed consisted of a mix of emotional qualities.[138] This, in many respects, has more in common with Aristotle's famous discussion of virtue being built on the prudent balancing of emotions in relation to differing circumstances. As Aristotle states in his *Ethics*: 'The man who gets angry at the right things and with the right people, and also in the right way and at the right time and for the right length of time, is commended'.[139] This perhaps constitutes what we would today refer to as 'emotional intelligence': maintaining the right balance between emotion and reason. In this sense, Clausewitz probably had more in common with Aristotle than a Stoic such as Seneca, but he certainly would have sympathised with the latter's warnings about the capacity for emotions to 'sweep us on with a force of their own and allow no turning back'.[140]

For the commander in war, the intelligence required is not one simply of logical reason, but one of emotional empathy and judgement that is capable of reading the likely emotional reactions of the enemy, one's own forces and others. Emotion in war is thus something that is not only to be actively exploited or tamed, but also a force to be sensitively judged and perceived. The use of force impacts not only on those directly caught up in war – the combatants themselves – but also on various external actors, all of whom may act so as to greatly transform the political or military landscape. For instance, many conquerors have failed because they did not account for the intense emotional reaction of a people deprived of their liberty. Judging the likely reaction of various 'agglomerations of emotion' is an inherently difficult and complex matter, but those who attempt to understand them and the myriad ways they impact on war stand a much better chance of success. These ideas might also relate to the individual soldier's effectiveness. In her book *Stoic Warriors*, Nancy Sherman advocates a 'moderate and mild stoicism' combining aspects of emotion and reason.[141] She warns about the debilitating effects of rage and anger against which the Stoics propose *apatheia* and fighting without anger,

137 Lucius Annaeus Seneca, 'On Anger', in Lucius Annaeus Seneca, *Moral and Political Essays*, ed. John M. Cooper and J. F. Procopé (Cambridge: Cambridge University Press, 1998), p. 25.

138 *On War*, I, 3, pp. 115–31.

139 Aristotle, *Ethics*, trans. J. A. K. Thomson (London: Penguin, 1976), p. 160.

140 Seneca, 'On Anger', p. 25.

141 Sherman, *Stoic Warriors*, p. 12.

but Sherman also demonstrates how some forms of emotional anger should be cautiously embraced.[142]

Thus, this tendency of the trinity is not something about which commanders should necessarily despair. The impact of emotions on war is ever-present, but strategically ambivalent and context dependent. The character and extent of their particular manifestations, whether in the political leadership, the fighting forces or 'the people', will depend on prevailing conditions. It is for the statesman, the commander and the soldier – in their respective spheres of competency – to read those conditions and, at the opportune moment, seek to tap, promote, shape, control or suppress all the various expressions of hate, anger, revenge, compassion, ambition, envy, pity and love that war's human actors invariably display. Passion and irrationality are forces that cannot be ignored or wished away.

*

This tendency has its roots in human nature and as such is an inescapable feature of human activity; a feature apt to be displayed at its greatest magnitude in the context of war. The use of force in human relations stirs emotions, the consequences of which can greatly impinge on the course of war. Those emotions can be understood in both an individualistic and collective sense. They are of a protean nature, capable of manifestation in many diverse forms that history records in abundance. Passion is not necessarily manifested as a malign force, nor one causing a rise to extremes at all times. Passion can equally be a force acting against war and the escalation of force, as evidenced in powerful pacifist sentiment. Passion is a universal aspect of war, but is not necessarily always dominant. Also, the precise dynamics of passion may be hard to pin down, being as it is an intangible, largely ephemeral force. Nevertheless, it is a very real element of war, particularly in its consequences, which are highly unpredictable and liable to violent fluctuations and intense expression.

If passion is not simply the enemy of sensible policy in relation to war, it is certainly a force the effects of which policy cannot escape and that sometimes the latter is hard put to control. Those responsible for making policy must either seek to mitigate passion's worst effects, cultivate its beneficial aspects or harness its potential violent energy in a controlled fashion. Otherwise, hopeless wars will be launched and the bonds of honour between warriors will wither while the waves of hatred sweep away the walls of reasoned behaviour in a flood of massacre, savagery and excess.

This concludes our analysis of Clausewitz's three primary tendencies. We are now in a position to provide a description of the trinitarian framework as a whole and this will be the subject of the following chapter.

---

142 Sherman, *Stoic Warriors*, p. 67 and p. 71.

Chapter 7
Interactions: The Trinity as a Unity

Now we have a clearer idea of the three separate primary elements that comprise the trinity, it is possible to consider how the trinity operates as a whole. This holistic perspective is crucial. Each separate tendency of the trinity does not stand alone and so, the previous three chapters, which have treated each tendency in isolation, are largely meaningless until they are brought together and their intended dynamic unity explained. Certain valuable insights can be taken from an analysis of each, but only when the structure of the whole is understood does Clausewitz's conception of the nature of war become clear.

We must also seek to understand how the primary tendencies relate, in theoretical terms, to the secondary trinity and war's multiple contexts at the tertiary level. These perspectives can help lessen the potential for misinterpretation. Not until the powerful play of contextual conditions are fully incorporated does its timelessness become clear. If the analysis of the separate tendencies appeared somewhat partial or incomplete, this is to be expected. Also, it should have become apparent that it is almost impossible to discuss one element without simultaneously considering the others: the true meaning of each tendency only fully emerges when considered together.

Prominent themes of analysis at the primary level

The previous three chapters, regarding the separate primary tendencies of the trinity, reveal a number of central themes that can be said to characterise all three; by this we mean the common attributes that pertain to each. It is important to recap these issues as a basis for understanding how the trinity operates as a whole.

Unilateral and multilateral perspectives

The trinity essentially presents a picture of war from the perspective of one actor, one 'side', perhaps even an alliance. Thus, some have claimed that in order to gain a complete perspective we have to visualise what we might term a 'clash of trinities'; that we need to bring together the trinities of each belligerent and analyse their interaction. This approach is not without its merits, yet appears to be largely derived from conceptions of the secondary level (as people/army/government) and fails to grasp the inherent complexity of the primary tendencies. The problem arises because, of course, opposing belligerents cannot share a common leadership or armed forces and they will relate primarily to their own distinct populations.

Yet, when considering the trinity at the primary level, as Clausewitz intended, it becomes clear, although it is not initially obvious, that the interactive nature of war is embodied in it.

None of the primary tendencies make sense without the existence of an 'other'. In each, unilateral perspectives are embedded in, and crucially dependent in terms of their meaning, on a multilateral reality. A purely unilateral perspective is certainly possible to a degree. So, policy is independently formulated based on the aims of the group. Important elements of uncertainty are generated internally or result from non-human external factors such as the weather, whereas the creative application of means can be directed toward lifeless objects. People can display violent emotions not necessarily caused by or directed at someone or something else. Yet, all these tendencies do not properly relate to war, or attain the specific form which Clausewitz intends, unless force is a feature of the interactive relationship *between* human groups.

Policy is meaningless without consideration of the broader multilateral political context within which it is inextricably embedded and influenced by. In Clausewitz's work the two perspectives almost at times elide because he sees them as being so indissolubly interconnected. The course of any war is shaped not only by what one or the other side hopes to achieve through force, but by the complex interplay and contortions created by the interaction. Many of the most important aspects of chance and uncertainty derive from such interaction. For instance, the difficulties of accurately assessing the enemy's intentions and capabilities are exacerbated by enemy efforts at deception, as well as the unpredictable occurrences when two independent forces interact. Arguably the most important source of passion in war derives from the interactive use of force whereby – as so powerfully evinced on the Eastern Front during World War II – hostility breeds hostility, fear of the other heightens group solidarity and is inflamed by dehumanising depictions of the opponent and the real and perceived brutalities of the enemy.

All three tendencies draw on one of Clausewitz's ontological foundations of the trinity: war's interactive nature. The nature of war is presented through the eyes of one actor, while simultaneously being embedded in and dependent on the context of interaction with other actors in conditions of violent conflict. The trinity is unilateral to the extent that it provides a route into an understanding of war from the angle of one political entity. The unilateral perspective is crucial for understanding war precisely because it is meaningless without being conceived of as comprised of unitary actors who come together in a situation of conflicting interests. While it may appear contradictory to describe the trinity as comprehensive given its unitary appearance, in actual fact, once its dependence on the multilateral, interactive conception is understood, the contradictions disappear.

Ambiguity in relation to the rise to extremes

Extremes in war can take many forms. The Thirty Years War, the Napoleonic Wars, The Taiping Rebellion and, in particular, the two World Wars perhaps encapsulate what we would accept as real-world examples. World War II claimed over 50 million lives, was fought over most of the globe, involved countless unspeakable horrors (such as the Holocaust and massacres of unarmed civilians) and saw the strategic bombing of entire cities. Other than the imagined nuclear Armageddon of a Third World War (after which, Einstein remarked, the Fourth will be waged with sticks and stones) there is no more clear instance of the appalling extremes war can reach.

Yet, we need to tread cautiously in our understanding of extremes, which to a considerable extent depends on our situation or the perspective from which we are observing a particular war. Consider for instance the modern popular and misleading concept of 'low-intensity warfare' used to designate conflicts short of major state warfare. As Mike Smith explains, such descriptions can be misleading when, for instance, the Vietnam War becomes subsumed under such a heading: 'it would be interesting to ask an American combat veteran whether he thought he was involved in a 'low-intensity war'?'[1] Just how extreme a conflict may be is highly subjective and even the search for means of measurement may prove misleading. So, for instance, it is estimated that America dropped three times as many bombs on Vietnam than were dropped by all combatants during World War II.[2] Indeed, Clausewitz's conception of extremes appears to have less to do with magnitude, scale or statistics and more to do with the character of the dynamics of the fighting in the context of the age.

Clausewitz – believing he had witnessed war which had 'attained the absolute in violence' – was keen to show how extremes could be explained, as well as why in other wars they were not always reached. His description of the rise to extremes in Book 1, Chapter 1 is an abstract discussion that is part of a dialectical argument. Clausewitz concludes that whether a rise to extremes actually occurs can only be determined by war in reality, embedded in its real political, social and physical contexts: it never simply follows the dictates of logic. This perspective is presented in the trinity, and, as a whole, is entirely ambiguous in relation to the rise to extremes. None of the three tendencies represent forces which necessarily cause escalation – they are *a priori* ambiguous in this respect. They all can lead to extremes, but equally they may exert countervailing and limiting forces.

Not only are the tendencies individually ambiguous in this sense, but the extent to which they will lead war one way or the other is dependent on a host of contextual factors which shape their manifestation in any particular situation. Perhaps, given the evidence of history, certain tendencies tend to gravitate more towards extremes – in particular, we might conclude that forces of passion and irrationality are generally more prone to cause extremes – but, nevertheless, all

1 Mike Smith, 'Strategy in an Age of "Low-Intensity" Warfare', in Isabelle Duyvesteyn and Jan Angstrom (eds), *Rethinking the Nature of War* (London: Frank Cass, 2005), p. 32.

2 Anthony Best, Jussi H. Hanhimaki, Joseph A. Maiolo, and Kirsten E. Schulze, *International History of the Twentieth Century* (London: Routledge, 2004), p. 294.

are conditional. We will generally observe great tension, both between and within the tendencies: some forces push war towards extremes, other exert a restraining influence. The trinity as a whole might usefully be conceived of as a 'vehicle', the direction of which is by no means preordained until the driver is in and moving.

Strategically neutral

From a subjective perspective, a close analysis of the elements of the trinity reveals their neutral quality. At first glance, for instance, it may appear that the tendency of chance and uncertainty – embodying as it does such concepts as friction – exerts an unambiguously negative and debilitating effect on strategic performance. This would be mistaken. While Clausewitz certainly wanted to reveal how, objectively, certain factors serve to impede all aspects of military activity (like movement in a resistant element), the strategic perspective reveals a different story. Precisely because friction is common to all belligerents, great advantages can accrue to those who are able to mitigate its worst effects – through inspired leadership, effective intelligence collection, combat experience and high morale – and exacerbate it for the enemy – through surprise or deception. From the strategic perspective, the realm of chance and uncertainty is one of great opportunity, which people of genius can exploit through daring stratagems or considered restraint.

Likewise, policy is neutral in terms of strategic performance. As in any purpose-means relationship, the purpose can exert a negative affect. To say that war is controlled by policy is not to say that policy is always wise. For instance, policy can be unclear, make unrealistic demands on the military or may itself be mistaken.[3] Furthermore, political interests may impinge on military decision-making or the interpretation of intelligence, all of which may be inimical to strategic success. A common recent concern, linked to possibilities created by new technologies, has been excessive and damaging political interference in military operations: the problem of the 'long screwdriver'. On the other hand, sound policy can have a hugely positive impact on strategic effectiveness – in particular, clear objectives provide the military with a firm idea of the kind of military operations they need to undertake.

Lastly, passion is also neutral in its strategic implications. On the one hand, it can lead to wholly unrealistic polices or operations driven by excessive ambition, blinding religious conviction or burning vengeance. Moral indignation might encourage humanitarian interventions which make little strategic sense.[4] Emotion

---

3 Policy can have an adverse affect if 'statesmen look to certain military moves and actions to produce effects that are foreign to their nature'. Carl von Clausewitz, *On War*, trans. and ed. Michael Howard and Peter Paret (New York: Everyman's Library, 1993), VIII, 6B, p. 735.

4 Colin S. Gray, 'The RMA and Intervention: A Sceptical View', in Colin McInnes and Nicholas J. Wheeler (eds), *Dimensions in Western Military Intervention* (London: Frank Cass, 2002), pp. 52–65.

can encourage the overuse of force, whereby levels of civilian casualties or brutalities turn opinion against the war and cause means to come detached from political ends. But, as Clausewitz knew, such passions could be a most effective strategic asset if exploited effectively by leaders. Also, he drew attention to other noble and positive emotions in war, which serve to create bonds of comradeship amongst soldiers and engender heightened displays of altruism and self-sacrifice. Such cohesive forces within armed forces can greatly contribute to fighting effectiveness and strategic success, when so much in war, as Clausewitz stressed, is dependent on psychological factors.

Context and implications at the secondary level

Although Clausewitz never employed the terms 'secondary' or 'subjective' in relation to the trinity they are useful ways of considering the level at which the primary trinity is embodied or expressed in real-world actors or 'subjects'. Jan Angstrom usefully describes the secondary level as the 'operationalisation'[5] of the primary trinity. The subjects Clausewitz identifies are, respectively, the government, the people and the commander and his army. It is quite clear from the text that Clausewitz utilises this secondary trinity as an illustrative device to instantiate the more ephemeral forces captured in the primary trinity through examples that would have made sense to his contemporaries and perhaps still do to us.[6] Paradoxically, this attempt at clarification through illustration has contributed to a great deal of misunderstanding; yet this is misunderstanding derived from a failure to read Clausewitz carefully. Some modern commentators have sought to couple each tendency too rigidly to each of these social groups or simply ignore the primary trinity altogether. While, for instance, Angstrom is right to point out that the secondary trinity 'was clearly influenced by the political and military context in which Clausewitz lived'[7] this need not prevent us from seeking a more flexible interpretation and certainly one not confined to the modern state.

A more flexible interpretation is facilitated by firmly recognising the role of context in the framework. In Chapter 3 we argued that this 'hidden' level of the trinity is continually acting on and within the social actors that conduct war. As the various conditions within which war is fought change, actors will inevitably be shaped by the prevailing circumstances which they themselves partly constitute. So, for instance, the cultural dimension of context will shape a group's leadership form and policy goals, the character of its fighting forces and the means it employs, as well as the attributes of the population and the way it responds to

5 Jan Angstrom, 'Introduction: Debating the Nature of Modern War', in Duyvesteyn and Angstrom, *Rethinking the Nature of War,* p. 5.

6 Christopher Bassford and Edward J. Villacres, 'Reclaiming the Clausewitzian Trinity', *Parameters*, Autumn, 1995.

7 Angstrom, 'Introduction', p. 5.

developments. Emphasising the importance of context was Clausewitz's primary means of overcoming the apparent dichotomy between the universal and particular, continuity and change in war. When we acknowledge the ever-present influence of context, we see how there is a constant interplay up and down the three levels of the trinity. The primary level of the trinity distils those timeless elements of war, which are manifested in a multitude of disparate and varied actors throughout history. Those actors conduct their own particular form of war (secondary level) moulded by changing conditions (tertiary, contextual level). It is a framework that reveals the universals in war, while allowing for great and continuous change.

The trinity and the state

In an important article written in 1995, Bassford and Edward Villacres made clear how a number of studies had employed the 'trinity' in an incorrect manner. As they noted, the proclivity to mistakenly describe the Clausewitzian trinity as comprised of the government, the military and the people was evident in the work of commentators both supportive and critical of Clausewitz's ideas.[8] Unfortunately, there has been no let-up in such studies adopting this formula and presenting it as the essence of Clausewitz's trinity. This particular representation of the secondary level is indeed a theoretical trinity, and one that can be of great value, but it is not ultimately a *Clausewitzian* trinity.

The apparent source of such interpretations appears to have been the highly influential book written by Harry Summers in 1982, which was a penetrating critique of the American Army's experience in Vietnam.[9] Summers's analysis benefited greatly from his use of the government-military-people distinctions as it allowed him to explore some of the important ways in which relations between these social entities broke down during the war. Summers's work presented a conception of the trinity that was tied to the modern state and its prominent social institutions, and this conception subsequently dominated popular interpretations of the trinity. In particular, when Creveld and Keegan embarked on their extensive critiques of Clausewitz's ideas, they did so on the basis of a derivative conception confined to the secondary level and, crucially, the modern state.

Similarly, these were highly influential works that caught the imagination of many and further consolidated the view that the secondary, people-army-government trinity represented *the* Clausewitzian trinity. Creveld's observations on the apparent demise of the state and the rise of forms of war in which the distinction between, for instance, the people and the military became almost meaningless appeared to be supported by events in the early 1990s, which lent his analysis credibility and, with it, his rejection of Clausewitz. This argument that Clausewitz is historically limited because, both in the past and increasingly

8 Bassford and Villacres, 'Reclaiming'.

9 Harry Summers, *On Strategy: The Vietnam War in Context* (Hawaii: University Press of the Pacific, 2003).

today, war was not sociologically delineated according to recognisable bodies of government, military and people betrays these authors' reliance on the secondary trinity, but also overstates the extent to which these divisions pertained in Clausewitz's time and the emphasis he placed on them in theory. Whether Creveld was right to announce the imminent demise of the state is an argument we will consider below, but it is sufficient to note that these studies influenced further generations of scholars who adopted this portrayal of the trinity.[10]

Thus, for many, *On War* became synonymous with old forms of state war and, so, as internecine, non-state warfare became seemingly more prevalent after 1989, this entailed that Clausewitz had to be discarded and new approaches unveiled. So Kaldor could state that, 'As the centralised, territorialised modern state gives way to new types of polity emerging out of new global processes, so war, as we presently conceive it [through Clausewitz], is becoming an anachronism'.[11] Much of the analysis contained in such works is of great value for students attempting to come to grips with the complexity of modern conflicts, but the interpretations of Clausewitz on which they proceed are inaccurate. Instead of returning to and exploring the implications of the primary trinity itself, they base their interpretations on the secondary trinity alone which significantly denies the flexibility, wider applicability and relevance of Clausewitz's theory.

When Clausewitz wrote *On War*, he no doubt had the modern state at the forefront of his mind: one of the most noticeable features of the history of the centuries prior to his era was the unstoppable rise of the modern state – primarily through its war-making capability – as the pre-eminent form of social organisation, particularly in Europe.[12] But, Clausewitz's conception of war is far more flexible than some commentators presume and can comfortably accommodate political entities other than the state. Kaldor is mistaken to assert that Clausewitz's definition 'implied that 'we' and 'our opponent' were states, and the 'will' of one state could be clearly defined. Hence war … is war between states for a definable political end, i.e. state interest'.[13] As Echevarria notes, 'his example of the Tartar tribes illustrates the case for nonstates, and puts paid to the mistaken notion that Clausewitz thought only in terms of the nationstate model'.[14] Political relationships between organised armed groups do not suddenly cease because the state is not involved. It is mistaken to claim that contests over relative power between self-interested groups disappear when the state is not at the centre of the confrontation.

10 Martin van Creveld, 'The Fate of the State', *Parameters*, Spring 1996, pp. 4–18.

11 Mary Kaldor, *New and Old Wars: Organized Violence in a Global Era* (Cambridge: Polity, 2002), p. 15.

12 Michael Howard, *The Causes of War* (Cambridge, Massachusetts: Harvard University Press, 1983), pp. 151–68.

13 Kaldor, *Old and New Wars*, p. 15.

14 Antulio J. Echevarria II, *Clausewitz and Contemporary War* (Oxford: Oxford University Press, 2007), p. 89.

Even a prominent critic of Clausewitz accepts that the trinity is 'easily adaptable to forms of warring social organisations that do not form states'.[15]

Yet, regardless of the fact that critics were aiming at an illusory target, we might well question just how limited the state conception actually is. The dominant agent in world politics today is the state and prognostications of its demise are perhaps too hasty. It appears that, as Isabelle Duyvesteyn has convincingly argued (despite problems in her own interpretation of Clausewitz),[16] commentators such as Creveld have adopted somewhat mistaken perspectives on conflicts which emerged during the 1990s. While they are correct to note the prevalence of intra-state conflict, Duyvesteyn reveals that most of those wars were actually more often than not explicitly *about* the state: belligerents fought in order to wrest control of the state.[17] Also, Creveld's belief that interstate conflicts are in decline is mistaken given that historically they have been rare events, while their ratio in relation to intrastate wars has been relatively stable for some time.[18] Nevertheless, again, the trinity is not dependent on the state perspective, although this is perhaps its most potent manifestation. As Gray remarks, 'Recognisable "war" predated, and will postdate, the modern states' system'.[19]

Secondary level analogues

So, the association of the trinity with the modern state was derived largely from a misreading of the secondary level in terms of its purpose and place in Clausewitz's theory. A theory which focuses on the government, army and people is naturally associated with the state and precludes examination of other forms of conflict. Three central arguments need to be presented in response to this mistaken conception. First, in a broad sense, Clausewitz was aware of wars that were not fought by modern states. Second, the secondary trinity as an illustrative device should not be confused with the key primary trinity. Third, accepting these two arguments, it is not difficult to conceive of 'analogues' of the three secondary subjects. We will focus on the third of these points.

Regardless of the apparent complexity of any war, analogues of the secondary trinity will be apparent in some shape or form. Indeed, reality is often undeniably

---

15 Jan Willem Honig, 'Strategy in a Post-Clausewitzian Setting', in Gert de Nooy (ed.), *The Clausewitzian Dictum and the Future of Western Military Strategy* (The Hague: Kluwer Law International, 1997), p. 110.

16 Her conception of Clausewitz is almost entirely confined to the secondary level. Isabelle Duyvesteyn, *Clausewitz and African War: Politics and Strategy in Liberia and Somalia* (London: Frank Cass, 2005), p. 1.

17 Duyvesteyn, *African War*, p. 108.

18 Colin McInnes, *Spectator-Sport Warfare: The West and Contemporary Conflict* (Colorado: Lynne Rienner, 2002), p. 58.

19 Colin S. Gray, 'Clausewitz Rules, OK? The Future is the Past – With GPS', *Review of International Studies*, Vol. 25, 1999, p. 169.

messy. The secondary subjective trinity can adopt many potential forms, but war cannot make sense without some grasp of the basic groups that conduct it or impact on its course. Bassford has convincingly argued that the use of more inclusive terms might help avoid misinterpretations. He suggests a more appropriate term for the 'government' might be 'leadership'; 'fighters' might usefully replace 'commander and army'; and 'popular base' can substitute for 'the people'.[20] Gow alternatively suggests 'political leadership, armed force and political community'.[21] This approach is to be welcomed as it is likely that analogues of these groups will be evident in most instances of war. As Honig states, 'Any community will have its leaders, fighters and common people'.[22] Importantly, the secondary level does not present a rigid sociological description of war in reality: as war's contexts and circumstances change, so too will it. Clausewitz's conception of war's subordination to policy, for instance, is applicable to any form of political entity which establishes goals and possesses the means to employ violence to attain its ends.

'Mainly' – translation between the primary and secondary levels

Aside from problems stemming from rigid, state-centric conceptions of the trinity, another connected problem at the secondary level relates to the way in which the primary tendencies translate or correspond to the secondary level. In the previous three chapters we analysed each tendency and in so doing revealed the way in which they cross many of the institutional boundaries illustrated by Clausewitz at the secondary level. This has often been overlooked and a strict correlation proposed. So, for instance, passion is associated solely with the people (or its analogue). Misinterpretation of this point has stemmed from a too literal reading of the second paragraph of the trinity, while ignoring the enormous implications of one word in the second paragraph: 'mainly' or 'mostly' (depending on the translation). Clausewitz states – in relation to passion, chance and politics – that, 'The first of these three aspects *mainly* concerns the people; the second the commander and his army; the third the government'.[23]

Again, this does not mean that the people, army, government formula will no longer apply – indeed, it may often be highly applicable – but that our conception of the trinity at the secondary level must be subservient to the primary meaning and the circumstances of the particular events being studied. If we determine that

20 Christopher Bassford, 'The Primacy of Policy and the 'Trinity' in Clausewitz's Mature Thought', in Hew Strachan and Andreas Herberg-Rothe, *Clausewitz in the Twenty-First Century* (Oxford: Oxford University Press, 2007), p. 82.

21 James Gow, 'The New Clausewitz? War, Force, Art and Utility – Rupert Smith on 21st Century Strategy, Operations and Tactics in a Comprehensive Context', *The Journal of Strategic Studies*, Vol. 29, No. 6, December 2006, p. 1168.

22 Honig, 'Strategy', p. 110.

23 *On War*, I, 1, § 28, p. 101.

the formula is relevant or correct, that should only be because our analysis revealed the primary tendencies operating along the lines of such social distinctions. How the trinity should be understood in this respect is made clear by Bassford and Villacres when they state that the secondary delineations should not be viewed as fixed attributes because 'each of the three [primary] categories … affects all of these human actors to some quite variable extent'.[24] The trinity morphs in its subjective manifestation in different times and places because war is 'more than a true chameleon that slightly adapts its characteristics to the given case'.[25] Returning briefly to the elements of the primary trinity underlines how they can span the sociological divides of people, army and government (or their more inclusive analogues).

Clausewitz associated the primary tendency of policy mainly with the government. This was for him a quite natural association and certainly still makes sense today, yet it is not difficult to conceive of a more inclusive conception. In a way not fully apparent to Clausewitz, the spread of democratisation during the nineteenth and twentieth centuries has greatly expanded the role of the people in policy, both in directly shaping it and as a factor governments must account for in decisions over the use of force. We should also note the important role the military often plays in shaping policy choices, sometimes actually usurping the reigns of power.[26] All actions in war (even at the lowest levels) have the potential to impact on political power dynamics in their various forms (consider, for instance, the massive political fallout from the Abu Ghraib scandal in Iraq resulting from the actions of individual soldiers).[27] The idea of soldiers being invested with considerable potential political agency is conveyed in the recent notion of the 'strategic corporal'.[28]

In many respects, it was quite right for Clausewitz to associate chance, uncertainty and genius most closely with the military sphere of the commander and his army. However, it is clear that the relevance of this tendency at the secondary level applies to other social groups. For Clausewitz, war as an act of policy is in many respects a chance undertaking, something of a gamble beyond the bounds of purely rational control, in which judgement and intuition are more appropriate than logical reason. As Ken Booth similarly observes, 'political behaviour is such a contingent activity that prediction can never be a science'.[29] Also, from an

24 Bassford and Villacres, 'Reclaiming'.

25 *On War*, p. 101.

26 See Hew Strachan, 'The Military in Politics', in Richard Holmes (ed.), *The Oxford Companion to Military History* (Oxford: Oxford University Press, 2001), pp. 723–30.

27 See Philippe Sands, *Lawless World: Making and Breaking Global Rules* (London: Penguin, 2006), p. 154.

28 See Charles C. Krulak, 'The Strategic Corporal: Leadership in the Three Block War', *Marines Magazine*, January 1999. Available online at: <http://www.au.af.mil/au/awc/awcgate/usmc/strategic_corporal.htm>, retrieved 3 June 2007.

29 Ken Booth, *Strategy and Ethnocentrism* (London: Croom Helm, 1979), p. 59.

objective perspective, the multilateral political context is one of the prime causes of war's unpredictability.

That a broader conception of this tendency at the secondary level of the trinity is required is evident when one considers the multidimensional and complex character of the modern battlespace, where overcoming chance and the use of creative talent is increasingly the preserve of actors other than simply the commander and his army. A wide array of organisations, aid agencies, reconstruction teams, diplomats and so on, all have a vital part to play in shaping political and military outcomes during and after conflict.[30]

In his description of the trinity Clausewitz states that passion 'mainly concerns the people' because 'the passions that are to be kindled in war must already be inherent in the people'.[31] As demonstrated in Chapter 6, this masks the extent to which the military and the political leadership are implicated in its meaning. Those responsible for formulating policy are equally susceptible to being motivated by ambition, dogmatism or ideological fanaticism. Military commanders and combatants of all descriptions are prone to be swept away in the heat of battle by feelings of fear, revenge, anger and hatred.

The point of this section has been to reveal that the forces in the primary trinity will be in operation regardless of the character or form of belligerents. Their manifestation will vary greatly at the secondary level, but their essence will remain the same. They will generally translate into the form of analogues of the Clausewitzian secondary trinity (political leadership, fighting forces and people), allowing for great differences in actual form in particular cases. Also, again, the primary elements cross sociological divides. So, for instance, some leaders combine both political and military roles, such as Alexander, Frederick or Napoleon. It is for analysts and strategists to correctly identify the character and dynamics of the social institutions engaged in war and to contemplate how the elements of policy, chance and passion are expressed through or manifested in them. Clausewitz simply provided us with the framework to begin such an analysis. Open interpretations, mindful of the emphasis placed on contextual factors, can accommodate the vast range of groups that have conducted war throughout the ages.

The trinity as an integrative and dynamic unity

As Paret states, Clausewitz believed that 'war was an activity in which each aspect influences and is influenced by others, and this interrelationship extended to the

30 See also Richard D. Hooker, 'Beyond *Vom Kriege*: The Character and Conduct of Modern War', *Parameters*, Summer 2005, p. 11.

31 *On War*, I, 1, § 28, p. 101.

social and political matrix of war'.[32] In Chapter 5, Book 2, Clausewitz states that 'in war, as in life generally, all parts of a whole are interconnected'.[33] The trinity is intended to reflect precisely this idea. The trinity is a unity and therefore must be comprehended as such. Importantly, this does not mean that the trinity is simply a combination of three elements placed side by side. It is much more than this and is intended to reflect the incredible complexity of war in reality.

It might be assumed from a basic presentation of the trinity that Clausewitz viewed war as a straightforward, ordered and structured activity; the three prominent elements of which can neatly be assessed independently. This interpretation is only further encouraged by a mistaken concentration on the secondary elements of the trinity, which suggests war displays a clear sociological division. This, as Beyerchen has demonstrated, would be a misleading reading of the trinity because 'the drive to comprehend the world through analysis, the effort to partition off pieces of the universe to make them amenable to study, opens the possibility of being blind-sided by the very artificiality of the partitioning practice'.[34] Clausewitz certainly partitioned reality this way, but he also wanted to stress that this was an artificial exercise and, as explained in the paragraphs following the substantive description of the trinity, one that masks incredible variation, interconnectedness and dynamism in reality.

A deeper exploration certainly reveals that the tendencies cannot be isolated in any neat manner – the boundaries between them are indistinct. Rather, central to Clausewitz's conception was the idea that 'the nature of war is complex and changeable'.[35] As Strachan has noted, 'Unlike many of the other triads which litter *On War* … this trinity really is three elements united into one'.[36] The importance of a theory capable of portraying this idea is evident in Clausewitz's notes, written at around the time he began to write what would become *On War*:

> Years of thinking about war, much association with able men who knew war, and a good deal of personal experience with it, have left certain ideas and convictions, and these he has preferred to present in compressed form, like small nuggets of pure metal … Perhaps a greater mind will soon appear to replace the individual nuggets with a single whole, cast of solid metal, free from all impurity.[37]

32 Peter Paret, *Clausewitz and the State: The Man, His Theories, and His Times* (Princeton: Princeton University Press, 1985), p. 189.

33 *On War*, II, 5, p. 184.

34 Alan Beyerchen, 'Clausewitz, Nonlinearity and the Importance of Imagery'. Available online at: <http://www.clausewitz.com/readings/Beyerchen/BeyerschenNon linearity2.pdf>, retrieved 20 April 2007.

35 *On War*, I, 2, p. 102.

36 Hew Strachan, *Carl von Clausewitz's On War: A Biography* (New York: Atlantic Monthly Press, 2007), p. 178.

37 Clausewitz, 'Author's Preface', in *On War*, p. 70.

Absent any other 'greater mind' it was left to Clausewitz himself to attempt this. The same desire is stated at the outset of *On War*: 'I propose to consider first the various elements of the subject, next its various parts or sections, and finally the whole in its internal structure'.[38] That whole is represented in the trinity and there are a number of vital points that need to be explained in order to comprehend the trinity holistically, as a unity, as an integrative and complete concept.

The idea of the search for the unity of phenomena is clearly evident in the culture of the German Movement. Its philosophical embodiment – that of German Idealism – was a powerful and dominant force at the time Clausewitz thought and wrote.[39] The idealists emphasised the 'integrative interrelation of all phenomena'.[40] Karl Ameriks describes how the idealists held that philosophy should be a deeply unified and autonomous enterprise, which ambitiously seeks to identify the structures that allow for a general account of 'how all experience, history, and nature hang together'.[41] Common to most – in particular Fichte and Hegel – was a dialectical mode of reasoning, employed to transcend apparent contradictions and arrive at a 'synthesis', based on the notion that through overcoming the oppositions inherent in concepts we can ascend to a more accurate understanding of reality.

The trinity, although not strictly Hegelian, can be seen as the theoretical means of expressing the unity of the phenomenon of war – Clausewitz's 'final synthesis' – which he arrived at after conducting an extensive dialectical examination of its central components. As Gat notes, while many of the ideas were in place to affect the resolution of the various inconsistencies in his theory, it was not until 1827, and the height of Hegel's influence in Berlin, that Clausewitz truly realised how best to proceed. Following idealist philosophy, in which all 'contradictions of reality were actually but differing aspects of a single whole',[42] he was able to expose war in its totality.

Importantly, and reflecting the influence of Idealism, all three aspects of the trinity serve to help define the others. None of them exist independently and all are in continual tension and interaction with one another: no single tendency has complete meaning separate from the others and no one tendency can be excluded at any time. As Clausewitz states, a theory which attempted to 'fix an arbitrary relationship between them would conflict with reality to such an extent that for this reason alone it would be totally useless'.[43] Their relationship to one another is by no means static, rather their relationship is one of dynamic interaction and

38 *On War*, I, 1, § 1, p. 83.

39 Roger Scruton, *A Short History of Modern Philosophy* (London: Routledge, 2005), p. 168.

40 Paret, *Clausewitz*, p. 150.

41 Karl Ameriks, 'Introduction: Interpreting German Idealism', in Karl Ameriks (ed.), *The Cambridge Companion to German Idealism* (Cambridge: Cambridge Unversity Press, 2006), pp. 4–10.

42 Gat, *Military Thought*, p. 234.

43 *On War*, I, 1, § 28, p. 101.

variation over time. The extent to which any one tendency is manifest at any one time can vary greatly, but none are ever completely absent.

It is vital to stress this point in relation to secondary trinity. Even if we conclude that in a certain war, the people are almost completely absent (as has been argued was largely the case in the 'cabinet wars' of the eighteenth century), this does not imply that passion and emotion will be too. This is because, as explained above, the primary tendencies are not exclusive to the groups with which Clausewitz 'mainly' identified them. Passion will be observable in both the leadership and the military in various ways and to varying extents. This is little more than a common-sense observation, but one that is apparently denied by commentators who adhere to rigid and parochial interpretations.

So, each tendency serves in some ways to define the limits of the others. For instance, policy, as representative of rational and goal-directed behaviour in war, cannot be understood alone: war is also pervaded by great chance, uncertainty and friction, while inescapable emotions impact on behaviour. In this way, although still an important element of war, the limits of rationality are exposed by accepting the simultaneous play of such forces. In this respect, policy struggles to maintain its control over its instrument because effectively linking means and ends is hampered by the play of chance events, uncertainty derived from psychological factors and the effects of blinding passion. Equally, however, the tendencies are not necessarily always in *competition* with one another, but can be mutually supportive. So, effective policy may be supported by empathetic understanding, creative imagination or intuitive judgement, combat effectiveness enhanced by rage and anger and military genius composed of a subtle blend of intellect and temperament. This holistic perspective gives the lie to characterisations of Clausewitz as a 'pure rationalist' and underlines the importance of understanding *On War* through the trinity, for it is the only point at which all his ideas find their true force as a complex unity. No idea (not even his famous dictum 'war as a continuation of politics') can be understood alone or as independently superior.

From one war to the next, and within the same war, the individual tendencies will vary in their relative influence. For instance, at times, political factors will be to the fore, while passion is less apparent, or at other times the chance and uncertainty of the military encounter suffused with myriad emotions will dominate, while the influence of policy is seemingly effaced. In some cases, all three may be strongly apparent. The point is to stress the infinite possibilities involved. This *dynamic variable interplay* determines the specific characteristics of any war and helps us explain why certain wars appear to display varying magnitudes of the tendencies. As Bassford has pointed out, the trinity conveys dynamism not balance or equilibrium.[44] The tendencies are in a continuous fluctuating, unstable and shifting relationship and, 'No one of these three elements is continually dominant,

44 Bassford, 'Primacy of Policy', p. 81.

not even policy'.[45] As Echevarria puts it, none of the tendencies 'are *a priori* more influential in determining the shape and course of actual conflict than any other'.[46]

Imagining the trinity – nonlinearity, complexity and chaos

Drawing these strands of analysis together, it is useful to draw on the insights and imagery of scientific nonlinearity and complexity theory in order to express the nature of the operation of war as encapsulated in the trinity. This is a perspective that has been encouraged by Beyerchen, who has stressed the importance of metaphor for understanding the concepts Clausewitz employs. He describes the trinity as a 'striking metaphor of nonlinearity'[47] and as Bassford notes, the trinity is a 'classic model of Chaos, in the modern scientific sense'.[48] So, what ideas does the science of nonlinearity convey? Essentially, it relates to complex, interactive and open systems, constitutive of many interacting variables, rather than of selective or simplistic relationships between a handful. Such systems display nonrecurring patterns of behaviour that are highly sensitive to the tiniest alterations in initial conditions. Slight changes in some variables can have wholly disproportionate effects leading to radically divergent outcomes. The overall result is a system typified by unpredictability, randomness and complexity.

The trinity in its basic form – perhaps reduced to such memorable formulas as 'politics, chance and passion' – can be extremely misleading for the student of war. We have already noted some of the reasons for this, such as the fact that each single tendency carries within it the weight of a great deal of theorising, detailed analysis and subsidiary concepts. Moreover, the image of the trinity provided by Clausewitz is 'not one of any kind of Euclidean triangle or triad, despite its understanding as such by many readers'[49] as misleadingly presented by a number of commentators.[50] This is precisely where the imagery provided by contemporary science of nonlinearity can be of value. It presents us with a means of grasping the intended dynamism of the relationship between the tendencies and of the operation of the trinity as a whole.

Immediately after presenting the substantive content of the trinity, Clausewitz notes that (in Bassford's translation) the 'task, therefore, is to keep our theory [of

45 Strachan, *Biography*, p. 178.

46 Echevarria, *Clausewitz*, p. 73.

47 Beyerchen, 'Nonlinearity'.

48 Bassford, 'Primacy of Policy', p. 81.

49 Beyerchen, 'Unpredictability', p. 70.

50 It is for this reason that employing diagrammatical representations of the trinity has been resisted in this book. Authors employing stylised depictions of the trinity include: Michael I. Handel, *Masters of War: Classical Strategic Thought* (Abingdon: Routledge, 2005) and, *War, Strategy and Intelligence* (London: Routledge, 1989); and Janeen Klinger. 'The Social Science of Carl von Clausewitz', *Parameters*, Spring 2006, pp. 86–7.

war] floating among these three tendencies, as among three points of attraction'.[51] As Beyerchen explains, 'when a magnet is released over three equidistant and equally powerful magnets, it moves irresolutely to and fro as it darts among the competing points of attraction'.[52] It is in this manner that war will chart a complex and unpredictable course amongst these three 'points of attraction'. Even though we may be able to anticipate the 'overall kind of pattern', any sort of 'quantitative predictability of the actual trajectory' of any war is impossible because even infinitely small variations in initial and subsequent conditions can cause a 'significantly different pattern'. War is an open system sensitive to differences in initial conditions and external influences, which will affect the course that the magnet charts between the various tendencies.[53]

This analogy with the non-linear phenomenon of the magnet certainly captures the incredible complexity of many wars, where there may be multiple actors interacting in shifting combinations. Such complexity can seriously hamper attempts at neat analysis. Nevertheless, the trinity was designed by Clausewitz with such complexity in mind – a fact which belies its seeming simplicity. It is the actual application of the trinity in the concrete case which remains the challenge for historians, analysts and strategists.

The relevance of the ideas of complexity theory might be particularly pertinent to many modern conflicts given the vast array of actors, operating in a globalised context, and enmeshed in vast and confusing transnational political webs. The unpredictable effects of an ever-present mass media filters perceptions and shapes, and is shaped by, popular attitudes. A growing complex of international legal constraints, military codes and diverse ethical preconceptions blurs the boundaries between acceptable and unacceptable behaviour. The proliferation of myriad futuristic technologies and high-tech weapons systems necessitates constant adaptation by combatants to this rapidly shifting material environment.

It should be noted that these insights do not amount to a cause for despair, either for the commander or for the student of war. Nor is the use of such imagery meant to suggest that in war all is chaos: there are of course some prominent linearities and predictable elements in war. Accepting the complexity of war is simply a further step towards properly understanding it. The trinity is an important means of feeling our way through that complexity. It condenses an enormously difficult subject into manageable elements that, if grasped, can greatly enhance knowledge. Moreover, the benefits of such nonlinear imagery in the strategic realm might be to encourage the vital attributes of adaptability, flexibility, responsiveness and innovation in military institutions.

The failure to conceive of the trinity in this way may reflect a natural human inclination for the reassuring presence of structure, order and predictability in human social affairs – something likely to be sought in military establishments

51 Bassford, 'Primacy of Policy', p. 77.

52 Beyerchen, 'Unpredictability', p. 70.

53 Beyerchen, 'Unpredictability', pp. 69–70.

as neat models promise prediction and greater control over intractable situations pervaded by fear and apprehension. As Steven Mann has argued, 'our view of reality rests on scientific paradigms' and the dominant outlook is shaped by the physical sciences … with their emphasis on deterministic and linear sequences of cause and effect.[54] Undoubtedly, Clausewitz's theory employs terminology from the physical sciences (centre of gravity, friction, etc). But, 'linear systems are often restrictive, narrow and brittle'[55] and can cause serious paralysis and confusion when developments and events do not fit neatly into the existing model. Beyerchen neatly sums up the potential weakness of linear conceptions:

> We need for our own sake to understand the limitations our imagination places upon us. Linearity … offers a narrow window on most natural and social systems. That narrowness sets blinders on our perception of reality and offers a weakness for an opponent to exploit … [while] an expanded sense of the complexity of reality can help us be more successfully adaptive amid changing circumstances.[56]

The trinity, in its basic form, potentially misconveys the enormous complexity of the ideas and concepts that it rests on. To paraphrase Clausewitz: everything in the theory of war is very simple, but the simplest thing is incredibly complex. It would be very easy to fall into the trap of believing that the trinity offers a useful short-cut to an advanced understanding of war. Undoubtedly, Clausewitz's theory *does* save the student of war a great deal of effort (to the extent that we do not have to start from first principles), yet this perspective ignores the underlying complexity both of his theory and of the nature of the relationships between and within the elements of the trinity.

*

So, to conclude, this chapter has sought to draw together the various strands of analysis presented in previous chapters and to reveal the operation of the trinitarian framework as a whole. It has demonstrated that there are certain levels and characteristics of the trinity that must be properly understood if an accurate comprehension of the meaning and explanative power of the framework is to be grasped.

The three primary tendencies taken together present a picture of war from the perspective of the individual belligerent, yet one that is crucially dependent on war's multilateral context. All three are also inherently ambiguous both in terms of whether they cause a rise to extremes or the extent to which they are strategically

---

54 Steven Mann, 'Chaos Theory and Strategic Thought', *Parameters*, Autumn 1992, pp. 54–5.

55 Beyerchen, 'Nonlinearity'.

56 Beyerchen, 'Nonlinearity'.

beneficial or harmful. Concrete statements in those respects can only be made according to precise manifestations in unique circumstances. The primary objective tendencies are expressed through history in their variable subjective forms – of a political leadership, fighting forces and 'the people' – and while certain primary elements may relate mostly to certain social groups, the connections are by no means rigid or deterministic.

Finally, we emphasised the intended *unity* of the trinity: no one tendency of the trinity can be understood alone, because war will always display all three elements simultaneously as they interact. The result is a conception of war that is dynamic, nonlinear, complex and inherently unpredictable. War is always comprised of an interplay between policy/politics, chance and passion – in all their internal complexity – and the relationship between them is constantly shifting in a random and uncontrollable manner.

Chapter 8
Reflections

In the form of his famous trinity, Clausewitz laid the basis for a comprehensive understanding of war. The trinity can perhaps be more mystifying than edifying, particularly if read in isolation and without an appreciation of his wider work; yet, it can serve a number of purposes for students of war. Much that has been written about *On War* has missed the point of the trinity. There are good reasons for this. In its existing form the text is imperfect, incomplete and at times perplexing to the modern mind. An early death prevented Clausewitz from fully developing his ideas. Perhaps, had he lived, his work would be less prone to basic misunderstandings, but we should be cautious in that regard. Misunderstandings derive from a failure to read the text carefully or to comprehend its central ideas in their proper historical context.

Moreover, it has to be doubted whether Clausewitz would ever have been finally satisfied with his work. In accordance with his overarching dialectical approach to the subject, he wrote 'organically', constantly reworking concepts and testing propositions in the light of new experiences or historical evidence.[1] These observations, rather than being redolent of a confused mind, reveal a thinker who had grasped the truth about his subject: no final word, no pithy theory would ever truly capture the essence of the phenomenon. This is why, as Gray maintains, we should treat *On War* 'as a living document'.[2]

If read with an open mind the trinity is capable of helping us to understand the essential dynamics of conflict even today. It cannot, of course, tell us anything substantive about modern conflict, but rather encourages consideration of the essential dynamics that underlie any situation of organised violence waged for political ends. The value of this lies in the way that essentials are often lost in the welter of overpowering images and the inordinate confusion of the 'here and now'. The trinity forces us to separate what at first sight appears utterly chaotic and appreciate the fundamental dynamics of war. Reality *is* often chaotic, but equally, in historical perspective, there is an underlying consistency to such complexity; if not 'method behind the madness', then perhaps continuity behind the confusion.

Importantly, the trinity does not stand alone, but must be understood as encompassing countless other ideas in Clausewitz's work. As Bassford and Villacres state, 'it would be a mistake to approach the trinity concept as a discrete,

---

1 Hew Strachan, *Carl von Clausewitz's On War: A Biography* (New York: Atlantic Monthly Press, 2007), p. 79.

2 Colin S. Gray, 'Clausewitz, History, and the Future Strategic World', *The Occasional*, No. 47, June 2003, p. 14.

bite-sized nugget of wisdom that can somehow be extracted from the larger work'.[3] The trinity is intended to nourish the intellect through a process which, as Clausewitz puts it, 'accustoms the mind to these truths'.[4] It is not meant to be employed as a rigid framework, whereby one's task would simply be to categorise facts and events into three distinct areas of reality and consider their relationship. Clausewitz makes this point himself, in relation to practice, when he states that, 'these truths are better served by a commander who has absorbed their meaning in his mind rather than one who treats them as rigid external rules'.[5]

The trinitarian framework

The trinity rests on three central suppositions, which were not explicitly stated in the section outlining the trinity, but that are central to Clausewitz's conception of war; the trinity cannot be properly understood without appreciating their ever-present underlying influence – to paraphrase Clausewitz, they are like the colour an artists gives to the underpainting which determines the tone of the canvas.

First, Clausewitz's understanding of war places humans with all their quirks, vanities, fears and hopes at the centre of the phenomenon. Second, throughout, Clausewitz emphasises 'the duel', the fact that war is always a 'continuous interaction of opposites'[6] between two or more belligerents – it is not an 'exercise of the will directed at inanimate matter', but at an 'animate object that *reacts*'.[7] Third, fighting and combat lie at the heart of the phenomenon, no matter how rarely violent engagements actually occur or even if force is only threatened.

Although seemingly commonsensical, these foundations are often marginalised or ignored. Military institutions have a tendency to seek answers to problems in technological elixirs, diverting attention away from human realities. This is particularly the case with respect to modern Western democracies where the existence of casualty averse publics has prompted the search for 'risk-free warfare' through the application of stand-off, precision weapons. Some militaries get so caught up in perfecting their own operations that they forget there is always a living and thinking enemy continually attempting to frustrate their plans. From that basic fact so much of war's paradoxical nature derives.[8] The belief that wars can be won without fighting, perhaps through political positioning, clever manoeuvre

3 Christopher Bassford and Edward J. Villacres, 'Reclaiming the Clausewitzian Trinity', *Parameters*, Autumn, 1995.

4 Carl von Clausewitz, *On War*, trans. and ed. Michael Howard and Peter Paret (New York: Everyman's Library, 1993), II, 5, p. 181.

5 *On War*, II, 5, p. 196.

6 *On War*, II, 2, p. 157.

7 *On War*, II, 3, pp. 173–4.

8 Edward Luttwak, *Strategy: The Logic of War and Peace* (Cambridge, Massachusetts: The Belknap Press, 2001).

or shrewd stratagems hides the fact that such successes are dependent on the threat or prospect of actual engagement: outcomes in war are ultimately decided by calculations in relation to the consequences – whether political or material – of the use of force. War, in Clausewitz's conception, is predicated on these foundations. Yet, in order to arrive at an accurate theory, he required methods that would enable him to avoid the pitfalls into which he believed earlier thinkers had fallen.

Rejecting both the stereotypically dogmatic and mechanistic Enlightenment approach and the exaggerated and fatalistic excesses of the Romantic outlook, Clausewitz struck a middling course. Rather than seeking any fixed doctrine or final conclusions, he forcefully expressed the limits of any theory of war. Nevertheless, he believed that by employing clear concepts free of jargon and approaching the subject in the spirit of scientific endeavour, sound theory could elucidate war's internal structures. Clausewitz was particularly conscious of ensuring that his theory remained anchored to reality through the testing of logical propositions against both history and personal experience. A direct consequence of this self-imposed criterion was the recognition that war is inescapably intertwined with psychological factors and effects. While incorporation of such factors only made the theorist's task more difficult, any theory that ignored them would conflict with reality to such an extent as to be rendered useless.

Moving beyond the depiction of war as either art or science, Clausewitz maintained that war is 'part of man's social existence' and is most akin to that from which it emerges: politics. Most importantly, he strove to bridge the divide between universal and particularist approaches, aware that the former leads to empty generalities and the latter to rigid doctrinal principles rapidly rendered obsolete. The trinity is the fruit of his theoretical labours, the final synthesis arrived at after years of intense study and historical analysis. The three primary tendencies represent those timeless elements of war, which, while varying in their relative importance over time, and from one war to the next, will be apparent in any war, in any age. So, what were those universal elements?

First, Clausewitz drew attention to the rational thread that runs through war in the form of the overarching influence of policy, whereby belligerents seek to tailor their available resources and means to ultimate ends through appropriate strategies and tactics. The influence of policy feeds into calculations regarding the costs, both material and psychological, that groups are willing to pay given the value of the war's object and the effort required. Although policy is never 'tyrant', can change dramatically as circumstances develop and may not always be deemed 'reasonable' according to universal standards of law or morality, the subordination of war to policy emphasises that no military victory is complete in and of itself.

But there was another less obvious, yet fundamental implication of war's subordination to policy; that war is inescapably intertwined with the wider realm of politics, within which policy is but one element. He held that war cannot be understood as an autonomous phenomenon, but was the continuation of *politics*, understood as an endless, multifaceted and dynamic process. Politics was the 'womb of war' and thus the main lines that run through war are fundamentally political. Yet, this wider perspective exerted a paradoxical effect, which emphasised to Clausewitz the limits of policy control on war. He understood that war, far from simply being a rational instrument, is also a continuation of the chaotic and unpredictable realm of the political, which introduces conflicting priorities and is beyond the total control of any one group.

Second, war is pervaded by chance and probability. The sources of this uncertainty lie primarily in war's interactive nature, the physical conditions within which war takes place and the human condition. War is always conducted against an enemy with an independent will, that never passively submits and constantly seeks to outwit, deceive and destroy. Chance happenings pervade war, either as a result of the surprise moves of the enemy, the impact of external factors such as the weather or the limits of the human mind to foresee all potential possibilities. The unexpected moves, both political and military, of third parties add to this chronically unpredictable environment. The uncertainty that results is exacerbated by the countless intangible factors involved and the inescapable ethnocentric blinkers that cloud objective analysis of enemy intentions and capabilities.

These are factors that even sophisticated technology cannot negate; information and intelligence in war is inevitably partial and prone to exaggeration and political manipulation. Combined with the effects of relentless danger, fear and physical and mental exertion, these factors coalesce to form a 'general friction' which makes war so much more difficult than it may appear on paper. Yet this realm of uncertainty is not necessarily cause for despair. One should remember that the enemy can no more easily escape its clutches and, more positively, it is an environment of great opportunities in which the creative and bold genius, directing skilled and determined troops, can flourish.

Third, contrary to the way in which he has often been represented, Clausewitz stressed the ever-present potential for irrational and emotional behaviour in war. Whenever force is present in human relations, the feelings cannot fail to be involved, while any putative level of civilisation is no safeguard against capture by such feelings, which are common to all humans and especially prone to intense expression during conflict. Yet, throughout his work, he recognises other potential sources of irrationality and it becomes clear, on close study, that these forces affect all actors in war. Ultimately, war is an activity in which rational behaviour can very easily be overwhelmed by the emotions. Furthermore, emotions in war are not only of a nefarious kind. Virtuous emotions such as intense love, altruism and self-sacrifice can exist and a surprising level of mutual respect and empathy may obtain between combatants. All such 'moral forces' can variably impinge on military effectiveness.

All three primary tendencies contain some common attributes. They each present a picture of war from the subjective point of view of an individual belligerent in war, yet the interactive and multilateral nature of war is inherently implied in each case. All three tendencies are essentially ambiguous in relation to the rise to extremes. Also, in strategic terms the three tendencies are neither necessarily positive or negative influences, but are dependent on the way in which they are handled or exploited by belligerents.

The three tendencies manifest themselves – at the secondary level – in different ways in different times and places due to the varying conditions in which societies conduct their wars. It is a framework that consciously accounts for the often rapidly changing conditions – captured through what we have termed the tertiary level of context – in which wars are fought and that greatly impact on the character of belligerents. The secondary subjective trinity is intended as a means of capturing the manifestation of the three primary forces in relation to particular historical wars. The primary tendencies of policy, chance and passion are typically associated, respectively, with the political leadership, fighting forces and popular base, but a closer examination reveals the varying impact of all three on all social groups.

Crucially, the trinity has to be conceived of as a unity – as an indivisible and integrated whole – in so far as none of the primary tendencies make sense in isolation and none is *a priori* more important than the others. As Clausewitz's final synthesis, the trinity is intended to convey the inextricable interaction of the central forces operating in war. Moreover, this unity – of three things in one – is not of a static form, but one that is endlessly morphing and fluctuating, displaying dynamism and constant change. Clausewitz employs the nonlinear scientific metaphor of a weight suspended over three magnets to convey the sense in which war charts a complex, unpredictable and random course between these three points of attraction. Thus, we can consider the whole framework as expressing a complex system of dialectical and ceaselessly shifting relationships between the three central forces as manifested in the particular belligerents involved, whose specific form is shaped by historical conditions, which are themselves constantly in flux.

As soon as one delves deeper into the separate elements of the trinity it becomes almost immediately apparent that any strict delineation between them is necessarily arbitrary and an insufficient reflection of the complexity of reality. Just as no human can be neatly compartmentalised into rational, creative and emotional sides, nor can war be so divided. The trinity is an incredible simplification of reality, yet a necessary one for understanding a phenomenon that can so easily escape the bounds of our mental capacity.

As a number of commentators have noted, there are a great many topics associated with war that Clausewitz ostensibly omits. *On War* is short on technology, economics, law or naval matters. Yet, equally of course, he says nothing about air combat, nuclear weapons or cyber-warfare. This reminds us that Clausewitz could not have written about *everything* to do with war, nor was that his intention. The trinity establishes a framework of underlying *essentials* and is designed to be flexible enough to accommodate fundamental change in war's manifestations in the concrete case, as shaped by the continual interplay of changing conditions, while remaining firmly anchored to the fundamental forces that drive all forms of human conflict: politics, chance and passion.

For any of those forces to be absent, war must necessarily become something else, for these elements are integral to its very nature. They radically alter in their expression, relative influence and relationship to one another from one war to the next and within the same war; at times certain elements may dominate, while others are pushed into the shade. Nevertheless, a theory that ignored any one of them 'would conflict with reality to such an extent that for this reason alone it would be totally useless'.[9] Even for commanders who believe war can be approached as an autonomous activity and attempt to achieve victory for victory's sake, war can never, in reality, be isolated from its political context. No matter how technologically advanced the armed force or how professional its intelligence agencies, when war is considered as a whole, there are too many contingent factors, intervening variables, intangibles and unknowns for uncertainty or chance to be ever truly banished. War is fought by humans and where violence enters into social relations, passions are inevitably stirred – whether virtuous or sinister – and this is not to mention the countless other sources of irrational behaviour.

Clausewitz today

Almost all discussion on the nature of war and strategy commences, wittingly and unwittingly, against the backdrop of purportedly Clausewitzian ideas and utilises concepts first fully developed by him. The flood of studies in recent years claiming there is something decidedly 'new' about contemporary war – whether due to the effects of globalisation, the media revolution or new forms of insurgency and terrorism – has led many commentators to return to Clausewitz and, ultimately, to the trinity.

Generally, it appears that the claim of novelty in contemporary war tends to rest less on objective observations of what went before, than on the ideas of Clausewitz, who is often painted as the chief interpreter of a dying form of inter-state war. Likewise, those who claim that war has not changed do so on the grounds that he essentially got it right. The central problem here lies not in the fact that Clausewitz is utilised as a starting point for such discussion, but that his ideas are often distorted in order to add weight to particular arguments. If Clausewitz represents all that purportedly went before, then any thesis claiming a new era of warfare can only benefit from knocking him off his pedestal. So, greater sense can be achieved in the wider debate on the nature of war if we better understand the theory on which contested claims are made. This means returning to the trinity, extrapolating its central conceptual components and their meaning and suggesting ways in which they allow for much greater flexibility of interpretation than is commonly allowed.

Yet, crucially, the trinity certainly cannot explain everything. Analysts of contemporary conflict – of individual wars, of conflicts in certain geographical

9 *On War*, I, 1, § 28, p. 101.

regions, or of its myriad forms, from guerrilla to major warfare – provide students with detailed descriptions of the events, causes and dynamics that define such conflicts. Such specialist studies are indispensable and Clausewitz cannot supplant them. They reveal the extremely complex character of wars that have emerged in recent decades; wars that appear radically different from traditional interstate wars prominent throughout most of the modern era.

Whether all the features of these wars are as novel as some proclaim need not worry us greatly; the point being that historical experience has undoubtedly revealed radically different forms of war. Clausewitz wrote at a time dominated by wars fought between the armies of sovereign states. That he witnessed and even lectured on emergent forms of guerrilla conflict and 'people's war' cannot hide this fact. His historical studies provided him with an awareness of different types of war, but the emphasis of his work was on interstate war and, more specifically, its Napoleonic manifestation. He knew nothing of the United Nations, mass media or laser-guided munitions: he was no Nostradamus. Essentially, the central problem is the one that Clausewitz himself grappled with: how to reconcile the universal and the particular.

Clausewitz's theoretical approach emphasised the limits of what he believed possible in any theory of war. He understood better than any existing thinkers the profound difficulty of making truly timeless observations on a subject as vast and complex as war. This recognition of extreme particularity almost led him to the relativistic conclusion that no general statements in theory were possible – indeed, he stated that each age would have held to its own theory of war. Clausewitz would no doubt have sympathised with those scholars who highlight the unique aspects of contemporary conflict.

Yet, as we have seen, Clausewitz was driven by an intense intellectual commitment to determine whether general propositions were in fact possible. His rigorous theoretical standards provided him with the means of carrying out this endeavour. He was ever attempting to establish the bounds of generalisable insights without succumbing to observations of meaningless banality. To achieve such an undertaking he would require not only a detailed comprehension of the dynamics of particular wars, but also a sense of how wars differed throughout history. In a manner largely unprecedented in military theory, Clausewitz truly began the process of, as he put it, 'thinking about the subject for years on end and testing each conclusion against the history of war', delving beneath the surface of war in an attempt to distinguish the essential from the incidental, the timeless from the particular. The trinity is the culmination – albeit incomplete – of this endeavour.

The trinity is a framework that is intended to convey dynamism, flexibility, change and complexity. We should approach the trinity as Clausewitz intended us to: as a guide to judgement, not a prescriptive manual with which any war can easily be understood or, indeed, fought. This point cannot be overstated. Clausewitz was highly conscious of the fact that contextual change could render irrelevant any dogmatic or prescriptive theory.

*

For many, *On War* will raise as many questions as it answers. Much of it is difficult to master, particularly for the uninitiated. It should be approached with caution, but also with relish, confident that one will emerge from close study of it with a better understanding of war. To fully comprehend its ideas, some knowledge of the context in which it was written is valuable. Perhaps more important is an understanding of its unfinished status. Failure to accommodate these factors has led to misleading distortions of Clausewitz's ideas. Nevertheless, those willing to take time to understand these problems can find in *On War* powerful ideas which speak across the ages.

Perhaps the central reason for Clausewitz's timelessness lies in his perceptive insight into the basic realities of human nature, particularly as they are wont to be expressed in times of war. Along with such thinkers as Thucydides and Machiavelli, he achieved that difficult task of seeing through the blinding effect of the present into the fundamental continuities of the human social condition. Not all his insights achieved this standard. Clausewitz was not only a philosopher of war, but a soldier responding to the exigencies of his time, hence much of *On War* is inevitably time-bound.

Nevertheless, war remains, with qualification, true to Clausewitz's original trinitarian conception. As he taught us, while the 'forms' of war are diverse and changing, its 'spirit' is universal.[10] Although it is unarguable that the character of wars and the contexts within which they are fought have changed considerably since Clausewitz wrote, war's objective nature has not undergone any profound transformation. When considering the more extreme pronouncements of 'new wars' scholars, we should listen to Thucydides: one of his protagonists comments that, 'Any novelty in an argument deceives you at once, but when the argument is tried and proved you become unwilling to follow it; you look with suspicion on what is normal and are the slaves of every paradox that comes your way'.[11]

The implications of the changing forms of war during the twenty-first century will no doubt continue to generate intense study and debate, but as Clausewitz taught, we should pause before being overawed by appearances and impressions. *On War* is not a textbook with all the answers about how we should think about war. Rather, it is an invitation to study an immensely complex phenomenon for which there are no quick and easy answers.

10 Gat, *Military Thought*, p. 205.

11 Thucydides, *Peloponnesian War*, p. 214.

Bibliography

Ameriks, Karl (ed.), *The Cambridge Companion to German Idealism* (Cambridge: Cambridge University Press, 2006).

Ameriks, Karl, 'Introduction: Interpreting German Idealism', in Karl Ameriks (ed.), *The Cambridge Companion to German Idealism* (Cambridge: Cambridge University Press, 2006), pp. 1–17.

Angstrom, Jan, 'Introduction: Debating the Nature of Modern War', in Isabelle Duyvesteyn and Jan Angstrom (eds), *Rethinking the Nature of War* (London: Frank Cass, 2005), pp. 1–27.

Aristotle, *Ethics*, trans. J. A. K. Thomson (London: Penguin, 1976).

Aristotle, *The Politics*, trans. T. A. Sinclair (London: Penguin, 1992).

Armstrong, David, Farrell, Theo, and Maiguashca, Bice (eds), *Force and Legitimacy in World Politics* (Cambridge: Cambridge University Press, 2005).

Aron, Raymond, 'The Evolution of Modern Strategic Thought', in Alistair Buchan (ed.), *Problems of Modern Strategy* (London: Chato and Windus, 1970), pp. 13–46.

Aron, Raymond, *Clausewitz: Philosopher of War*, trans. C. Booker and N. Stone (New Jersey: Prentice-Hall, 1985).

Aron, Raymond, *Peace and War* (New Jersey: Transaction, 2003).

Ashdown, Paddy, *Swords and Ploughshares: Bringing Peace to the 21st Century* (London: Weidenfeld & Nicolson, 2007).

Barnes, Jonathan, *Aristotle: A Very Short Introduction* (Oxford: Oxford University Press, 2000).

Bassford, Christopher, 'John Keegan and the Grand Tradition of Trashing Clausewitz', *War in History*, Vol. 1, No. 3, November 1994, pp. 319–36.

Bassford, Christopher, 'The Primacy of Policy and the "Trinity" in Clausewitz's Mature Thought', in Hew Strachan and Andreas Herberg-Rothe (eds), *Clausewitz in the Twenty-First Century* (Oxford: Oxford University Press, 2007), pp. 74–90.

Bassford, Christopher, 'Policy, Politics, War, and Military Strategy'. Available online at: <http://www.clausewitz.com/readings/Bassford/StrategyDraft/index.htm>, retrieved 15 January 2012.

Bassford, Christopher, 'Tip-Toe Through the Trinity or The Strange Persistence of Trinitarian Warfare'. Available online at: <http://www.clausewitz.com/readings/Bassford/Trinity/Trinity8.htm>, retrieved 20 January 2012.

Bassford, Christopher and Villacres, Edward J., 'Reclaiming the Clausewitzian Trinity', *Parameters*, Autumn 1995.

Bellamy, Christopher, *Knights in White Armour: The New Art of War and Peace* (London: Pimlico, 1998).

Bellamy, Christopher, *Absolute War, Soviet Russia in the Second World War: a Modern History* (London: Macmillan, 2007).

Best, Anthony, Hanhimaki, Jussi H., Maiolo, Joseph A., and Schulze, Kirsten E., *International History of the Twentieth Century* (London: Routledge, 2004).

Betts, Richard K., 'Analysis, War, and Decision: Why Intelligence Failures Are Inevitable', *World Politics*, Vol. 31, No. 1, October 1978, pp. 61–89.

Beyerchen, Alan, 'Clausewitz, Nonlinearity and the Unpredictability of War', *International Security*, Vol. 17, No. 3, Winter 1992, pp. 59–90.

Beyerchen, Alan, 'Clausewitz, Nonlinearity and the Importance of Imagery'. Available online at: <http://www.clausewitz.com/readings/Beyerchen/BeyerschenNonlinearity2.pdf>, retrieved 20 April 2007.

Biddle, Stephen, *Military Power: Explaining Victory and Defeat in Modern Battle* (Princeton: Princeton University Press, 2004).

Black, Jeremy, *War* (London: Continuum, 2001).

Black, Jeremy, *War and the New Disorder in the 21st Century* (London: Continuum, 2004).

Black, Jeremy, *War Since 1945* (London: Reaktion Books, 2004).

Black, Jeremy, *Introduction to Global Military History: 1775 to the Present Day* (Abingdon: Routledge, 2005).

Black, Jeremy, 'War and International Relations: a Military Historical Perspective on Force and Legitimacy', in David Armstrong, Theo Farrell, and Bice Maiguashca (eds), *Force and Legitimacy in World Politics* (Cambridge: Cambridge University Press, 2005), pp. 127–42.

Blanning, Tim, 'Epilogue: The Old Order Transformed 1789–1815', in Euan Cameron (ed.), *Early Modern Europe* (Oxford: Oxford University Press, 1999), pp. 345–74.

Blanning, Tim, *The Pursuit of Glory: Europe 1648–1815* (London: Penguin, 2007).

Bond, Brian, *The Pursuit of Victory: From Napoleon to Saddam Hussein* (Oxford: Oxford University Press, 2006).

Booth, Ken, *Strategy and Ethnocentrism* (London: Croom Helm, 1979).

Boucher, David, *Political Theories of International Relations* (Oxford: Oxford University Press, 1998).

Bourke, Joanna, *An Intimate History of Killing: Face-to-Face Killing in the Twentieth Century* (London: Granta Books, 1999).

Boyle, Nicholas, *German Literature: A Very Short Introduction* (Oxford: Oxford University Press, 2008).

Brodie, Bernard, *War and Politics* (London: Cassell & Company Ltd, 1974).

Brodie, Bernard, 'The Continuing Relevance of *On War*', in Carl von Clausewitz, *On War*, trans. and ed. Michael Howard and Peter Paret (New York: Everyman's Library, 1993), pp. 50–65.

Brodie, Bernard, 'A Guide to the Reading of *On War*', in Carl von Clausewitz, *On War*, trans. and ed. Michael Howard and Peter Paret (New York: Everyman's Library, 1993), pp. 775–853.

Brown, Chris, *Understanding International Relations* (Hampshire: Palgrave, 2001).

Brown, Chris, Nardin, Terry, and Rengger, Nicholas (eds), *International Relations in Political Thought: Texts from the Ancient Greeks to the First World War* (Cambridge: Cambridge University Press, 2003).

Browning, Peter, *The Changing Nature of Warfare: The Development of Land Warfare from 1792 to 1945* (Cambridge: Cambridge University Press, 2002).

Buchan, Alistair (ed.), *Problems of Modern Strategy* (London: Chato and Windus, 1970).

Bull, Hedley, 'Strategic Studies and its Critics', *World Politics*, Vol. 20, No. 4, July 1968, pp. 593–605.

Butterfield, Herbert and Wight, Martin (eds), *Diplomatic Investigations* (London: George Allen & Unwin, 1966).

Cameron, Euan (ed.), *Early Modern Europe* (Oxford: Oxford University Press, 1999).

Carlyle, Thomas, *Heroes and Hero Worship* (1928). Available online at: <http://www.online-literature.com/thomas-carlyle/heroes-and-hero-worship/2/>, retrieved 12 May 2008.

Carr, Edward Hallett, *What is History?* (London: Penguin, 1990).

Clark, Christopher, *Iron Kingdom: The Rise and Downfall of Prussia, 1600–1947* (London: Penguin, 2007).

Clark, Kenneth, *Civilisation: A Personal View* (St Ives: John Murray, 2005).

Clark, Wesley, *Winning Modern Wars: Iraq, Terrorism, and the American Empire* (New York: Public Affairs, 2004).

Clausewitz, Carl von, *On War*, trans. J. J. Graham and ed. F. N. Maude (London: Penguin, 1982).

Clausewitz, Carl von, *Historical and Political Writings*, ed. and trans. Peter Paret and Daniel Moran (New Jersey: Princeton University Press, 1992).

Clausewitz, Carl von, *On War*, trans. and ed. Michael Howard and Peter Paret (New York: Everyman's Library, 1993).

Clausewitz, Carl von, *The Principles of War*, trans. and ed. Hans W. Gatzke (New York: Dover Publications, 2003).

Cobban, Alfred, *A History of Modern France – Volume II: 1799–1871* (London: Penguin, 1965).

Coker, Christopher, *Waging War Without Warriors? The Changing Culture of Military Conflict* (London: Lynne Rienner, 2002).

Cornish, Paul, 'The United States and Counterinsurgency: "Political First, Political Last, Political Always"', *International Affairs*, Vol. 85, No. 1, January 2009, pp. 61–79.

Creveld, Martin van, *The Transformation of War* (New York: Free Press, 1991).

Creveld, Martin van, 'The Fate of the State', *Parameters*, Spring 1996, pp. 4–18.

Creveld, Martin van, *Command in War* (Cambridge, Massachusetts: Harvard University Press, 2002).

Creveld, Martin van, *The Art of War: War and Military Thought* (New York: Smithsonian Books, 2005).

Creveld, Martin van, 'The Eternal Clausewitz', in Michael I. Handel (ed.), *Clausewitz and Modern Strategy* (Abingdon: Routledge, 2005), pp. 35–50.

Creveld, Martin van, *The Culture of War* (New York: Ballantine Books, 2008).

Crozier, Brian, 'The Strategic Uses of Revolutionary War', in Alistair Buchan (ed.), *Problems of Modern Strategy* (London: Chato and Windus, 1970), pp. 198–219.

Deist, Wilhelm, 'The Road to Ideological War: Germany, 1918–1945', in Williamson Murray, MacGregor Knox, and Alvin Bernstein, (eds), *The Making of Strategy: Rulers, States and War* (Cambridge: Cambridge University Press, 2007), pp. 352–92.

Dockrill, Michael, 'The Experience of War', in Lawrence Freedman (ed.), *War* (Oxford: Oxford University Press, 1994), pp. 9–11.

Duyvesteyn, Isabelle, *Clausewitz and African War: Politics and Strategy in Liberia and Somalia* (London: Frank Cass, 2005).

Duyvesteyn, Isabelle and Angstrom, Jan (eds), *Rethinking the Nature of War* (London: Frank Cass, 2005).

Dyer, Gwynne, *War: The Lethal Custom* (New York: Carroll & Graf, 2005).

Echevarria II, Antulio J., *Clausewitz and Contemporary War* (Oxford: Oxford University Press, 2007).

Echevarria II, Antulio J., 'Clausewitz and the Nature of the War on Terror', in Hew Strachan and Andreas Herberg-Rothe (eds), *Clausewitz in the Twenty-First Century* (Oxford: Oxford University Press, 2007), pp. 196–218.

Ehrenreich, Barbara, *Blood Rites: A History of Passion in War* (St Ives: Virago, 1998).

Esdaile, Charles, *Napoleon's Wars* (London: Penguin, 2008).

Evans, Dylan, *Emotion: A Very Short Introduction* (Oxford: Oxford University Press, 2001).

Ferguson, Niall, *The War of the World: History's Age of Hatred* (London: Penguin, 2006).

Fleming, Colin M., 'New or Old Wars? Debating a Clausewitzian Future', *The Journal of Strategic Studies*, Vol. 32, No. 2, April 2009, pp. 213–41.

Forsyth, Murray, 'The Tradition of International Law', in Terry Nardin and David R. Mapel, *Traditions of International Ethics* (Cambridge: Cambridge University Press, 2002), pp. 23–41.

Freedman, Lawrence (ed.), *War* (Oxford: Oxford University Press, 1994).

Freedman, Lawrence, *The Revolution in Strategic Affairs*, Adelphi Paper 318 (London: Routledge, 1998).

Freedman, Lawrence, *The Evolution of Nuclear Strategy* (Hampshire: Palgrave Macmillan, 2003).

Freedman, Lawrence, 'The Age of Liberal Wars', in David Armstrong, Theo Farrell, and Bice Maiguashca (eds), *Force and Legitimacy in World Politics*, (London: Frank Cass, 2005), pp. 93–107.

Freedman, Lawrence, *The Transformation of Strategic Affairs*, Adelphi Paper 379 (London: Routledge, 2006).

Gat, Azar, 'Machiavelli and the Decline of the Classical Notion of the Lessons of History in the Study of War', *Military Affairs*, Vol. 52, No. 4, October 1998, pp. 203–5.

Gat, Azar, *A History of Military Thought: From the Enlightenment to the Cold War* (Oxford: Oxford University Press, 2001).

Gat, Azar, *War in Human Civilisation* (Oxford: Oxford University Press, 2006).

Gilbert, Felix, 'Machiavelli', in Peter Paret (ed.), *Makers of Modern Strategy: From Machiavelli to the Nuclear Age* (Oxford: Oxford University Press, 1986), pp. 11–31.

Goethe, Johann Wolfgang von, *The Sorrows of Young Werther* (London: Penguin, 1989).

Gow, James, 'The New Clausewitz? War, Force, Art and Utility – Rupert Smith on 21st Century Strategy, Operations and Tactics in a Comprehensive Context', *The Journal of Strategic Studies*, Vol. 29, No. 6, December 2006, pp. 1151–70.

Gray, Colin S, 'What RAND Hath Wrought', *Foreign Policy*, No. 4, Autumn 1971, pp. 111–29.

Gray, Colin S., *War, Peace, and Victory: Strategy and Statecraft for the Next Century* (New York: Simon and Schuster, 1990).

Gray, Colin S., 'On Strategic Performance', *Joint Forces Quarterly*, Winter 1995–6, pp. 30–6.

Gray, Colin S., 'War, Politics, and RMA – the Legacy of Clausewitz', *Joint Forces Quarterly*, Winter 1995–6, pp. 76–80.

Gray, Colin S., 'Clausewitz Rules, OK? The Future is the Past – With GPS', *Review of International Studies*, Vol. 25, 1999, pp. 161–82.

Gray, Colin S., *Modern Strategy* (Oxford: Oxford University Press, 1999).

Gray, Colin S., 'Why Strategy Is Difficult', *Joint Forces Quarterly*, Summer 1999, pp. 6–12.

Gray, Colin S., 'The RMA and Intervention: A Sceptical View', in Colin McInnes and Nicholas J. Wheeler (eds), *Dimensions in Western Military Intervention* (London: Frank Cass, 2002), pp. 52–65.

Gray, Colin S., 'Clausewitz, History, and the Future Strategic World', *The Occasional*, No. 47, June 2003, pp. 1–24.

Gray, Colin S., *Another Bloody Century: Future Warfare* (London: Weidenfeld & Nicolson, 2005).

Gray, Colin S., 'How Has War Changed Since the End of the Cold War?', *Parameters*, Spring 2005, pp. 14–26.

Gray, Colin S., *Recognizing and Understanding Revolutionary Change in Warfare: The Sovereignty of Context* (Carlisle, PA: Strategic Studies Institute, 2006).

Gray, Colin S., *Strategy and History: Essays on Theory and Practice* (Abingdon: Routledge, 2006).
Gray, Colin S., *War, Peace and International Relations: An Introduction to Strategic History* (Abingdon: Routledge, 2007).
Gray, John, *Black Mass: Apocalyptic Religion and the Death of Utopia* (London: Penguin, 2007).
Grimsley, Mark, 'Surviving Military Revolution: The U.S Civil War', in MacGregor Knox and Williamson Murray (eds), *The Dynamics of Military Revolution: 1300–2050* (New York: Cambridge University Press, 2008), pp. 74–91.
Guevara, Ernesto 'Che', *Guerrilla Warfare* (London: Souvenir Press, 2003).
Handel, Michael I., *War, Strategy and Intelligence* (London: Routledge, 1989).
Handel, Michael I., 'Clausewitz in the Age of Technology', in Michael I. Handel (ed.), *Clausewitz and Modern Strategy* (Abingdon: Routledge, 2005), pp. 51–92.
Handel, Michael I. (ed.), *Clausewitz and Modern Strategy* (Abingdon: Routledge, 2005).
Handel, Michael I., *Masters of War: Classical Strategic Thought* (Abingdon: Routledge, 2005).
Hegel, Georg, *Lectures on the Philosophy of History* (Whitefish: Kessinger, 2004).
Herberg-Rothe, Andreas, *Clausewitz's Puzzle: The Political Theory of War* (Oxford: Oxford University Press, 2007).
Herbig, Katherine L., 'Chance and Uncertainty in *On War*', in Michael I. Handel (ed.), *Clausewitz and Modern Strategy* (Abingdon: Routledge, 2005), pp. 95–116.
Herodotus, *The Histories*, trans. Aubrey de Sélincourt (London: Penguin, 1996).
Heuser, Beatrice, *Reading Clausewitz* (London: Pimlico, 2002).
Hobsbawm, Eric, *The Age of Revolution: Europe 1789–1848* (St Ives: Abacus, 1999).
Holmes, Richard (ed.), *The Oxford Companion to Military History* (Oxford: Oxford University Press, 2001).
Homer, *The Iliad*, trans. Robert Fagles (London: Penguin, 1991).
Honig, Jan Willem, 'Strategy in a Post-Clausewitzian Setting', in Gert de Nooy (ed.), *The Clausewitzian Dictum and the Future of Western Military Strategy* (The Hague: Kluwer Law International, 1997), pp. 109–21.
Hooker, Richard D., 'Beyond *Vom Kriege*: The Character and Conduct of Modern War', *Parameters*, Summer 2005, pp. 4–17.
Horne, Alistair, *A Savage War of Peace: Algeria 1954–1962* (New York: New York Review Books, 2006).
Howard, Michael, 'War as an Instrument of Policy', in Herbert Butterfield and Martin Wight (eds), *Diplomatic Investigations* (London: George Allen & Unwin, 1966), pp. 193–200.
Howard, Michael, *Studies in War and Peace* (London: Temple Smith, 1970).

Howard, Michael, *War in European History* (Oxford: Oxford University Press, 1976).

Howard, Michael, *The Causes of War* (Cambridge, Massachusetts: Harvard University Press, 1983).

Howard, Michael, 'The Influence of Clausewitz', in Carl von Clausewitz, *On War*, trans. and ed. Michael Howard and Peter Paret (New York: Everyman's Library, 1993), pp. 29–49.

Howard, Michael, *The Franco-Prussian War* (Abingdon: Routledge, 2000).

Howard, Michael, *The Invention of Peace and the Reinvention of War* (London: Profile Books, 2001).

Howard, Michael, *Clausewitz: A Very Short Introduction* (Oxford: Oxford University Press, 2002).

Howard, Michael, 'Military Power and International Relations', *International Affairs*, Vol. 85, No. 1, January 2009, pp. 145–55.

Howard, Michael, Andreopoulos, George, and Shulman, Mark (eds), *The Laws of War: Constraints on Warfare in the Western World* (New York: Yale University Press, 1994).

Ignatieff, Michael, *The Warrior's Honor: Ethnic War and the Modern Conscience* (London: Vintage, 1999).

Ignatieff, Michael, *Virtual War: Kosovo and Beyond* (London: Vintage, 2001).

Jervis, Robert, 'Hypotheses on Misperception', *World Politics*, Vol. 20, No. 3, April 1968, pp. 454–49.

Jomini, Antoine Henri de, *The Art of War*, trans. G. H. Mendell and W. P. Craighill (Texas: El Paso Norte Press, 2005).

Judah, Tim, *Kosovo: War and Revenge* (New Haven: Yale University Press, 2000).

Kaldor, Mary, *New and Old Wars: Organized Violence in a Global Era* (Cambridge: Polity, 2002).

Kaldor, Mary, 'Inconclusive Wars: Is Clausewitz Still Relevant in these Global Times', *Global Policy*, Vol. 1, Issue 3, October 2010, pp. 271–81.

Kant, Immanuel, *Political Writings*, ed. Hans Reiss (Cambridge: Cambridge University Press, 1993).

Kant, Immanuel, *Critique of Pure Reason* (Cambridge: Cambridge University Press, 1998).

Kaplan, Robert D., *Warrior Politics: Why Leadership Demands a Pagan Ethos* (New York: Vintage Books, 2002).

Karnow, Stanley, *Vietnam: A History* (London: Pimlico, 1994).

Keegan, John, *A History of Warfare* (London: Pimlico, 2004).

Keegan, John, *Intelligence in War* (London: Pimlico, 2004).

Keegan, John, *The Face of Battle* (London: Pimlico, 2004).

Keen, David, *The Economic Functions of Violence in Civil Wars*, Adelphi Paper 320 (Oxford: Oxford University Press, 1998).

Kennedy, Paul, *The Rise and Fall of the Great Powers* (London: Fontana Press, 1989).

Kirkpatrick, Jeane, *Democracy and Double Standards: Rationalism and Reason in Politics* (New York: Simon and Schuster, 1982).
Klinger, Janeen, 'The Social Science of Carl von Clausewitz', *Parameters*, Spring 2006, pp. 79–89.
Knox, MacGregor, 'Conclusion: Continuity and Revolution in the Making of Strategy', in Williamson Murray, MacGregor Knox, and Alvin Bernstein (eds), *The Making of Strategy: Rulers, States and War* (Cambridge: Cambridge University Press, 2007), pp. 614–45.
Knox, MacGregor and Murray, Williamson, *The Dynamics of Military Revolution: 1300–2050* (New York: Cambridge University Press, 2008).
Krepenevich, Andrew F., *The Army and Vietnam* (Baltimore: John Hopkins University Press, 1988).
Krulak, Charles C., 'The Strategic Corporal: Leadership in the Three Block War', *Marines Magazine*, January 1999. Available online at: <http://www.au.af.mil/au/awc/awcgate/usmc/strategic_corporal.htm>, retrieved 3 June 2007.
Lamport, F. J., 'Introduction', in Friedrich Schiller, *The Robbers and Wallenstein* (London: Penguin, 1979), pp. 7–19.
Lawrence, Thomas. E., 'The Evolution of a Revolt', *Army Quarterly and Defence Journal*, October 1920, pp. 55–69.
Liddell Hart, Basil, *Strategy* (New York: Meridian, 1991).
Livy, *The Early History of Rome*, trans. Aubrey de Sélincourt (London: Penguin, 2002).
Lloyd, Henry, *History of the Late War between the King of Prussia and the Empress of Germany and her Allies* (London: 1781).
Luttwak, Edward, *Strategy: The Logic of War and Peace* (Cambridge, Massachusetts: The Belknap Press, 2001).
Lynn, John A., 'Nations in Arms 1763–1815', in Geoffrey Parker (ed.), *The Cambridge History of Warfare* (New York: Cambridge University Press, 2005), pp. 189–216.
Lynn, John A., 'States in Conflict', in Geoffrey Parker (ed.), *The Cambridge History of Warfare* (New York: Cambridge University Press, 2005), pp. 167–88.
Machiavelli, Niccolò, *The Prince*, trans. George Bull (London: Penguin, 1999).
Machiavelli, Niccolò, *The Art of War*, trans. Ellis Farneworth (Cambridge, Massachusetts: Da Capo Press, 2001).
Machiavelli, Niccolò, *The Discourses*, trans. Leslie J. Walker (London: Penguin, 2003).
McInnes, Colin, *Spectator-Sport Warfare: The West and Contemporary Conflict* (Colorado: Lynne Rienner, 2002).
McInnes, Colin and Wheeler, Nicholas J. (eds), *Dimensions in Western Military Intervention* (London: Frank Cass, 2002).
McNeill, William H., *The Pursuit of Power: Technology, Armed Force, and Society* (Chicago: University of Chicago Press, 1984).

Manis, Jerome G. and Meltzer, Bernard N., 'Chance in Human Affairs', *Sociological Theory*, Vol. 12, No. 1, March 1994, pp. 45–56.

Mann, Steven, 'Chaos Theory and Strategic Thought', *Parameters*, Autumn, 1992, pp. 54–68.

Metz, Steven, 'A Wake for Clausewitz: Toward a Philosophy of 21st Century Warfare', *Parameters*, Winter 1994–5, pp. 126–32.

Moody, Peter R., 'Clausewitz and the Fading Dialectic of War', *World Politics*, Vol. 31, No. 3, April 1979, pp. 417–33.

Moran, Daniel, 'The Instrument: Clausewitz on Aims and Objectives in War', in Hew Strachan and Andreas Herberg-Rothe (eds), *Clausewitz in the Twenty-First Century* (Oxford: Oxford University Press, 2007), pp. 91–106.

Moran, Daniel, 'Strategic Theory and the History of War', Naval Postgraduate School. Available online at: <http://www.clausewitz.com/readings/Moran-StrategicTheory.pdf#zoom=100>, retrieved 20 January 2012.

Münkler, Herfreid, *The New Wars* (Cambridge: Polity, 2005).

Murray, Williamson, 'Thinking About Revolutions in Military Affairs', *Joint Forces Quarterly*, Summer 1997, pp. 69–76.

Murray, Williamson, 'Towards World War', in Geoffrey Parker (ed.), *The Cambridge History of Warfare* (New York: Cambridge University Press, 2005), pp. 249–77.

Murray, Williamson and Grimsley, Mark, 'Introduction: On Strategy', in Williamson Murray, MacGregor Knox, and Alvin Bernstein (eds), *The Making of Strategy: Rulers, States and War* (Cambridge: Cambridge University Press, 2007), pp. 1–23.

Murray, Williamson and Knox, MacGregor, *Dynamics of Military Revolution, 1300–2050* (Cambridge: Cambridge University Press, 2008).

Murray, Williamson., Knox, MacGregor., and Bernstein, Alvin (eds), *The Making of Strategy: Rulers, States and War* (Cambridge: Cambridge University Press, 2007).

Nagl, John A., *Learning to Eat Soup with a Knife: Counterinsurgency Lessons from Malaya and Vietnam* (Chicago: University of Chicago Press, 2005).

Nardin, Terry and Mapel, David R. (eds), *Traditions of International Ethics* (Cambridge: Cambridge University Press, 2002).

Nooy, Gert de (ed.), *The Clausewitzian Dictum and the Future of Western Military Strategy* (The Hague: Kluwer Law International, 1997).

Oakeshott, Michael, *Rationalism in Politics and Other Essays* (Indianapolis: Liberty Fund, 1991).

Overy, Richard, *Why the Allies Won* (London: Pimlico, 1996).

Owens, Bill, *Lifting the Fog of War* (Baltimore: Johns Hopkins University Press, 2001).

Palmer, R. R., 'Frederick the Great, Guibert, Bülow: From Dynastic to National War', in Peter Paret (ed.), *Makers of Modern Strategy: From Machiavelli to the Nuclear Age* (Oxford: Oxford University Press, 1986), pp. 91–119.

Paret, Peter, *Clausewitz and the State: The Man, His Theories, and His Times* (Princeton: Princeton University Press, 1985).

Paret, Peter, 'Clausewitz', in Peter Paret (ed.), *Makers of Modern Strategy: From Machiavelli to the Nuclear Age* (Oxford: Oxford University Press, 1986), pp. 186–213.

Paret, Peter (ed.), *Makers of Modern Strategy: From Machiavelli to the Nuclear Age* (Oxford: Oxford University Press, 1986).

Paret, Peter, 'The Genesis of *On War*', in Carl von Clausewitz, *On War*, trans. and ed. Michael Howard and Peter Paret (New York: Everyman's Library, 1993), pp. 3–28.

Paret, Peter, *Understanding War: Essays on Clausewitz and the History of Military Power* (New Jersey: Princeton University Press, 1993).

Paret, Peter and Moran, Daniel, 'Introduction to Part One', in Carl von Clausewitz, *Historical and Political Writings*, ed. and trans. Peter Paret and Daniel Moran (New Jersey: Princeton University Press, 1992), pp. 3–14.

Parker, Geoffrey (ed.), *The Cambridge History of Warfare* (New York: Cambridge University Press, 2005).

Parker, Geoffrey, 'Introduction: The Western Way of War', in Geoffrey Parker (ed.), *The Cambridge History of Warfare* (New York: Cambridge University Press, 2005), pp. 1–11.

Parkinson, Roger, *Clausewitz: A Biography* (New York: Stein and Day, 1979).

Perry, Marvin, *An Intellectual History of Modern Europe* (Boston: Houghton Mifflin Company, 1993).

Pinkard, Terry, *German Philosophy 1760–1860: The Legacy of Idealism* (Cambridge: Cambridge University Press, 2008).

Pitkin, Hanna F., *Fortune is a Woman* (Chicago: University of Chicago Press, 1999).

Polk, William, *Violent Politics: A History of Insurgency, Terrorism and Guerrilla War, from the American Revolution to Iraq* (New York: Harper Collins, 2007).

Rapoport, Anatol, 'Editor's Introduction', in Carl von Clausewitz, *On War*, trans. J. J. Graham and ed. F. N. Maude (London: Penguin, 1982), pp. 11–80.

Rengger, Nicholas, 'On the Just War Tradition in the Twenty-First Century', *International Affairs*, Vol. 78, No. 2, 2002, pp. 353–63.

Rengger, Nicholas, 'The Judgement of War: On the Idea of Legitimate Force in World Politics', in David Armstrong., Theo Farrell, and Bice Maiguashca (eds), *Force and Legitimacy in World Politics* (Cambridge: Cambridge University Press, 2005), pp. 143–61.

Roberts, John M., *The Hutchinson History of the World* (London: Hutchinson, 1976).

Rothenberg, Gunther E., 'The Age of Napoleon', in Michael Howard, George Andreopoulos, and Mark Shulman (eds), *The Laws of War: Constraints on Warfare in the Western World* (New York: Yale University Press, 1994), pp. 86–97.

Rothenberg, Gunther E., *The Napoleonic Wars* (London: Cassell, 2004).

Rousseau, Jean-Jacques, *A Discourse on Inequality* (London: Penguin, 1984).
Roxborough, Ian, 'Clausewitz and the Sociology of War', *The British Journal of Sociology*, Vol. 45, No. 4, December 1994, pp. 619–36.
Sands, Philippe, *Lawless World: Making and Breaking Global Rules* (London: Penguin, 2006).
Saxe, Maurice de, *Reveries on the Art of War*, trans. and ed. Thomas R. Phillips (New York: Dover Publications, 2007).
Schiller, Friedrich, *The Robbers and Wallenstein* (London: Penguin, 1979).
Scruton, Roger, *A Short History of Modern Philosophy* (London: Routledge, 2005).
Seneca, Lucius Annaeus, *Moral and Political Essays*, ed. John M. Cooper and J. F. Procopé (Cambridge: Cambridge University Press, 1998).
Shaw, Martin, *The New Western Way of War: Risk-Transfer War and its Crisis in Iraq* (Cambridge: Polity, 2005).
Sherman, Nancy, *Stoic Warriors: The Ancient Philosophy Behind the Military Mind* (New York: Oxford University Press, 2005).
Shy, John, 'Jomini', in Peter Paret (ed.), *Makers of Modern Strategy: From Machiavelli to the Nuclear Age* (Oxford: Oxford University Press, 1986), pp. 143–85.
Shy, John, 'The Cultural Approach to the History of War', *The Journal of Military History*, Vol. 57, October 1993, pp. 13–26.
Skinner, Quentin, *Machiavelli: A Very Short Introduction* (Oxford: Oxford University Press, 2000).
Smith, Hugh, *On Clausewitz* (Hampshire: Palgrave Macmillan, 2005).
Smith, Mike, 'Strategy in an Age of "Low-Intensity" Warfare: Why Clausewitz is Still More Relevant than his Critics', in Isabelle Duyvesteyn and Jan Angstrom, (eds), *Rethinking the Nature of War* (London: Frank Cass, 2005), pp. 28–64.
Smith, Rupert, *The Utility of Force: The Art of War in the Modern World* (London: Penguin, 2005).
Soanes, Catherine (ed.), *The Oxford Dictionary, Thesaurus, and Wordpower Guide* (Oxford: Oxford University Press, 2001).
Strachan, Hew, 'The Military in Politics', in Richard Holmes (ed.), *The Oxford Companion to Military History* (Oxford: Oxford University Press, 2001), pp. 723–30.
Strachan, Hew, *European Armies and the Conduct of War* (Kings Lynn: Routledge, 2006).
Strachan, Hew, *Carl von Clausewitz's On War: A Biography* (New York: Atlantic Monthly Press, 2007).
Strachan, Hew, 'Clausewitz and the Dialectics of War', in Hew Strachan and Andreas Herberg-Rothe, *Clausewitz in the Twenty-First Century* (Oxford: Oxford University Press, 2007), pp. 14–44.
Strachan, Hew and Herberg-Rothe, Andreas (eds), *Clausewitz in the Twenty-First Century* (Oxford: Oxford University Press, 2007).

Strachan, Hew and Herberg-Rothe, Andreas, 'Introduction', in Hew Strachan and Andreas Herberg-Rothe (eds), *Clausewitz in the Twenty-First Century* (Oxford: Oxford University Press, 2007), pp. 1–44.

Summers, Harry, *On Strategy: The Vietnam War in Context* (Hawaii: University Press of the Pacific, 2003).

Taber, Robert, *The War of the Flea: The Classic Study of Guerrilla Warfare* (Washington, DC: Potomac Books, 2002).

Thucydides, *History of the Peloponnesian War*, trans. Rex Warner (London: Penguin, 1972).

Tolstoy, Leo, *War and Peace*, trans. Rosemary Edmonds (London: Penguin, 1982).

Townshend, Charles, 'Counter-insurgency', in Richard Holmes (ed.), *The Oxford Companion to Military History* (Oxford: Oxford University Press, 2001), pp. 231–2.

Tzu, Sun, *The Art of War*, trans. T. Cleary (Boston: Shambhala, 1988).

Waldman, Thomas, 'Politics and War: Clausewitz's Paradoxical Equation', *Parameters*, Vol. 40, No. 3, Autumn 2010, pp. 48–60.

Waldman, Thomas, '"Shadows of Uncertainty": Clausewitz's Timeless Analysis of Chance in War', *Defence Studies*, Vol. 10, No. 3, September 2010, pp. 336–68.

Waldman, Thomas, 'Clausewitz and the Study of War', *Defence Studies*, Vol. 12, No. 3, June 2012, pp. 345–74.

Waltz, Kenneth N., *Man, the State, and War: a Theoretical Analysis* (New York: Columbia University Press, 2001).

Watts, Barry D., *Clausewitzian Friction and Future War* (Washington: National Defense University, 1996).

Wawro, Geoffrey, *Warfare and Society in Europe, 1792–1914* (London: Routledge, 2000).

Weber, Max, *Politics as a Vocation* (1919). Available online at: <http://www.ne.jp/asahi/moriyuki/abukuma/weber/lecture/politics_vocation.html>, retrieved 6 June 2008.

Wedgwood, Victoria, *The Thirty Years War* (New York: New York Review Books, 2005).

Wight, Martin, *Power Politics* (London: Pelican, 1979).

Winters, Harold A., *Battling the Elements: Weather and Terrain in the Conduct of War* (Baltimore: John Hopkins University Press, 1998).

Wood, Neal, 'Some Common Aspects of the Thought of Seneca and Machiavelli', *Renaissance Quarterly*, Vol. 21, No. 1, Spring 1968, pp. 11–23.

Wood, Neal, 'Introduction' in Niccolò Machiavelli, *The Art of War*, trans. Ellis Farneworth (Cambridge, Massachusetts: Da Capo Press, 2001), pp. ix–lxxxiii.

Index